CONTENTS

The Quantum Soul	1
A Personal Note from the Author	2
Chapter 1: The Mystery of Consciousness	5
Chapter 2: Quantum Physics – A Brief Overview	11
Chapter 3: The Nature of Reality – Classical vs. Quantum Views	19
Chapter 4: The Observer Effect and Consciousness	27
Chapter 5: The Holographic Universe and the Nature of Reality	35
Chapter 6: Quantum Soul – A Metaphysical Perspective	44
Chapter 7: Entanglement and the Interconnectedness of Souls	52
Chapter 8: Superposition and the Multidimensional Soul	60
Chapter 9: Non-Locality and the Boundless Nature of the Soul	68
Chapter 10: The Quantum Field and Universal Consciousness	76
Chapter 11: Near-Death Experiences and Quantum	84

Reality	
Chapter 12: Time, Eternity, and the Quantum Self	92
Chapter 13: The Cosmic Dance of Free Will and Destiny	100
Chapter 14: The Veil Between Worlds	107
Chapter 15: The Illusion of Separation and the Unity of Existence	115
Chapter 16: The Silent Symphony of Energy and Form	122
Chapter 17: The Infinite Mirror	129
Chapter 18: The Breath of Creation	136
Chapter 19: The Alchemy of Thought	143
Chapter 20: The Luminous Fabric of Memory	151
Chapter 21: The Veins of Infinity	159
Chapter 22: The Veil of Perception	166
Chapter 23: The Silent Pulse of Eternity	174
Chapter 24: The Alchemy of Light and Shadow	181
Chapter 25: The Whispering Thread of Destiny	188
Chapter 26: The Hidden Geometry of the Soul	195
Chapter 27: The Ethereal Web	202
Chapter 28: The Silent Seed of Transformation	209
Chapter 29: The Unfolding Horizon	216
Chapter 30: The Breath of the Eternal	223
Epilogue	230
References	236
Glossary of Terms	242
Acknowledgements	249

Copyright Information 251
Disclaimer 252

THE QUANTUM SOUL

The Metaphysics Of Consciousness

By Dr Bhaskar Bora

A PERSONAL NOTE FROM THE AUTHOR

My journey, once marked by certainty and driven by purpose, has transformed in ways I could never have anticipated. It is no longer about grand achievements or the pursuit of external success, but about the quiet, tender moments that reveal the true essence of life—moments of love, care, and presence. What you hold in your hands is not just a collection of words, but a testament to resilience, a story woven from the delicate threads of struggle, acceptance, and ultimately, renewal.

There was a time when my life flowed with the grace of a symphony, every note in perfect harmony. As a doctor, my days were filled with the pulse of life itself—offering hope, easing suffering, and healing with steady hands. The white coat I wore wasn't just a symbol of my profession; it embodied my very identity; an outward reflection of the healer I believed I was destined to be. The lives I touched, the people I helped—it all gave profound meaning to my existence.

But life, in its mysterious and unpredictable ways, had other plans. In one swift, unforeseen moment, the world I knew unravelled. First came the spinal cord injury, stripping away the physical strength I had relied upon. Then, the shadow of cancer darkened the horizon, a stark reminder of life's fragility. The world of medicine, where I once found so much joy and purpose, suddenly slipped away, leaving a vast emptiness in its wake—a silence where once there had been meaning.

Gone were the bustling corridors of the hospital, replaced by the quiet solitude of my home. No longer a "Doctor," I found myself standing at the edge of an uncertain future, my hands—once so steady with the knowledge of healing—trembling with questions I wasn't ready to face. Without the title, without the work that had defined me for so long, who was I? What was left of me when everything I had known was no longer within reach?

In that silence, in the stillness of a life interrupted, I began to uncover something unexpected. The role of a disabled husband and father, once a distant concept, became my new reality—one that held unexpected grace.

What began as an effort to nurture my relationships, to find solace in this new world, slowly evolved into a profound inward journey.

I found healing in the spiritual—a rhythm of meditation, reading, and reflection that allowed me to rediscover the parts of myself I thought were lost. As I immersed myself in books, audiobooks, and hours of research, I began to understand that this new chapter of my life was not an ending, but a rebirth. The solitude of these years, the quiet hours of writing and reflection, gave birth to the very pages you hold in your hands now.

It is with deep gratitude that I share these words with you, knowing that they carry with them not just knowledge, but a piece of my soul. I hope that these reflections and insights offer you a fresh perspective on life and perhaps some nourishment for your own journey.

That we cannot control what the universe throws at us but how we react to those curveballs define who we are and what we make of our lives.

CHAPTER 1: THE MYSTERY OF CONSCIOUSNESS

In the nebulous dance between the finite and the infinite, there exists a question so ancient, so unyielding in its complexity, that it has haunted the corridors of human thought since the dawn of our awareness. What is consciousness, this flickering flame within the dimness of the cosmos? Is it but an ephemeral glimmer born of the mind, or does it stretch beyond the prison of flesh, transcending the mortal veil into realms untrodden by the senses? To approach this enigma is to stand before a precipice where logic falters, and only the most daring of souls venture into the abyss of understanding.

Consciousness — a word spoken in hushed tones by philosophers and mystics, uttered in fragmented whispers by those who seek to decipher the eternal riddle of existence. It is the ineffable presence that lurks behind the gaze of every living creature, observing, perceiving, and yet remaining tantalisingly elusive. The ancients called it the essence of being, the breath of gods bestowed upon mortal clay, while the moderns dissect it, hoping to unravel its secrets with the scalpel of science. Yet, as we shall see, the mystery of consciousness is not so easily dismantled. It is a mirror reflecting infinity, and to grasp it fully is to confront the limits of thought itself.

From the earliest moments of recorded history, humans have marvelled at their awareness, at the vividness of experience that colours the mundane world with meaning. To feel, to think, to dream — these are the privileges of consciousness, gifts bestowed upon us without explanation. Why do we perceive? What animates this theatre of the mind where thoughts, sensations, and memories converge in a sublime choreography? For centuries, we have sought answers, but the farther we dig, the deeper the mystery becomes.

Some would argue that consciousness is nothing more than the flickering of neurons, a byproduct of the brain's intricate machinery. The materialists, confident in the supremacy of the physical world, propose that it is in the convoluted folds of grey matter where consciousness finds its birth. But even they, with their microscopes and equations, cannot escape the profound question that looms above all others: How does matter, inert and insensate, give rise to subjective experience? How does the brain, a physical organ like any other, conjure the vibrant world of thought, sensation, and self-reflection?

In their laboratories, scientists probe the mind, hoping to catch a glimpse of the spark that ignites consciousness. They speak of synapses, of chemical reactions and electrical impulses, and yet these mechanisms, however detailed, cannot explain the most basic fact of our existence — that we feel. To be conscious is not simply to process information like a machine; it is to be, to experience the richness of life, to inhabit a world of colour, sound, and emotion. No equation, no formula, can capture the essence of what it means to be aware.

The ancient sages, less inclined to the sterile precision of modern science, took a different path. They peered inward, into the depths of their own being, hoping to find within themselves the answers that elude the grasp of logic. They spoke of the soul, of a divine spark that animates the body and survives the grave. For them, consciousness was not a phenomenon to be explained but a mystery to be embraced. It was the very substance of existence, a luminous thread that connected all living things to the fabric of the cosmos.

Plato, the philosopher-poet of ancient Greece, offered one of the earliest and most enduring conceptions of consciousness. For him, the soul was the seat of all knowledge and perception, a vessel of eternal truth imprisoned in the body. The physical world, with its fleeting shadows and illusions, was but a pale reflection of the higher reality known only to the soul. To be conscious, for Plato, was to glimpse this higher realm, to remember the eternal forms from which all things derive. And yet, even he, in his wisdom, could not fully explain how the soul perceives, how it bridges the chasm between the eternal and the temporal.

Across the ancient world, in the mystic lands of the East, the Upanishadic sages spoke of Atman, the self that resides beyond the fleeting illusions of the world. In their vision, consciousness was not an individual phenomenon but a universal force, the very essence of the cosmos itself. To know one's consciousness was to know the universe, for the self and the world were one. The veil of ignorance, they taught, could be lifted through deep meditation, allowing the soul to merge with the infinite. Consciousness, in this view, was not something to be acquired or explained but something to be realised — an awakening to the true nature of existence.

In the halls of modern philosophy, the German idealists echoed these ancient ideas, speaking of consciousness as the foundation of all reality. For Immanuel Kant, the mind shaped the world as much as it perceived it, imposing its categories upon the raw data of experience. In his wake, thinkers like Hegel and Schelling proposed that consciousness was not merely an individual phenomenon but a cosmic force unfolding through history in a grand dialectic of self-realisation. The world, they argued, was the manifestation of spirit, and to understand consciousness was to understand the very nature of reality.

But if consciousness is the foundation of reality, then we must ask: What is its origin? How does it arise from the void, from the primordial chaos that existed before time and space? Here, we step into the domain of metaphysics, where the laws of physics give way to the strange and the sublime. The materialists, confident in their mastery of the physical world, would have us believe that consciousness is a mere byproduct of evolution, an accidental side effect of complex biology. But this explanation, elegant though it may seem, fails to capture the profound mystery of conscious experience.

For if we are but the products of blind forces, if

our thoughts and emotions are nothing more than the chemical reactions of a highly evolved organism, then how do we explain the richness of our inner worlds? How do we explain the beauty of a sunset, the sorrow of loss, the joy of love? These experiences are not reducible to matter; they are not mere functions of biology. They are something more, something transcendent, that defies the cold logic of materialism.

Some have sought to bridge the gap between matter and mind by invoking the strange and mysterious world of quantum mechanics. In the subatomic realm, where particles exist in states of superposition, where the observer alters the outcome of events simply by looking, there are hints of a deeper reality, one in which consciousness plays a fundamental role. Could it be that the mind, like the particles of quantum physics, exists in multiple states at once, that our awareness is not confined to the here and now but stretches across time and space?

It is a tantalizing possibility, one that has captured the imagination of both scientists and mystics alike. Theories abound, from the idea that consciousness is a quantum phenomenon, arising from the entanglement of particles in the brain, to the more radical notion that consciousness is the fundamental fabric of the universe itself. In this view, the physical world is but a projection of consciousness, a shadow cast upon the wall of the cosmic cave.

But even if we accept the possibility of a quantum consciousness, we are still left with the question of its nature. What is this awareness that watches from behind the veil of perception? Is it a unified force, a single mind that spans the cosmos, or is it fragmented and divided among the countless beings who inhabit the universe? Does consciousness arise only in living creatures, or does it permeate all of reality, from the tiniest particle to the vast expanse of the galaxies?

In our search for answers, we must be cautious not to lose ourselves in speculation. While the mystery of consciousness invites us to explore the farthest reaches of thought, it is also a deeply personal experience, one that can only be understood by turning inward. To know consciousness, we must not only study the stars and the particles that dance in the void; we must also listen to the quiet voice within, the whisper of the soul that speaks in moments of silence and solitude.

It is here, in the stillness of our awareness, that we may begin to glimpse the true nature of consciousness. It is not something to be grasped or dissected but something to be experienced lived, and felt. It is the silent witness to the unfolding drama of existence, the eternal observer who watches as the world spins and time flows. And perhaps, in the end, this is all we can ever truly know: that we are conscious, that we are here, and that within us lies a mystery deeper than the stars.

Consciousness is not a puzzle to be solved but a journey to be embraced. It is the path that leads us to the heart of the universe, to the very source of being itself. As we walk this path, we may come to realise that the greatest mysteries are not those we seek to unravel but those we carry within us, waiting to be awakened.

CHAPTER 2: QUANTUM PHYSICS – A BRIEF OVERVIEW

In the dim corridors of human inquiry, where thought and reason have long wandered in search of truth, there exists a threshold — one that neither the mystic's whisper nor the scientist's calculation has entirely grasped. Beyond it lies a domain woven not by the logic of Newtonian order but by the undulating rhythms of uncertainty and paradox, where particles become waves, and waves dissolve into nothingness. It is a world where the smallest fragments of matter, those elusive architects of existence, refuse to be governed by the steadfast rules that bind the larger universe. This is the realm of quantum physics, where the very fabric of reality bends and warps in ways both wondrous and bewildering.

To speak of quantum physics is to summon a lexicon of the arcane — particles, entanglements, probabilities, and superpositions — each term a mere reflection of the profound mystery that lurks beneath. But to truly understand its significance, one must first relinquish the comfort of certainty and abandon the notion that reality is a solid structure upon which we may rest our assumptions. Here, in this subatomic world, we encounter a place of flux, a liminal space where what is is always on the verge of becoming something else.

Long ago, when the cosmic veil first began to lift, humanity sought to organise the world into patterns, imposing order upon the seemingly chaotic dance of nature. From the stars that wheeled above to the smallest grains of sand, the ancient thinkers saw in the universe a grand clockwork, a precise mechanism in which each part moved in harmony with the rest. And for centuries, this vision held true. The laws of motion, gravity, and cause and effect ruled supreme, casting the universe as a vast, predictable machine — an infinite play of matter and force.

Then came the fracture. A quiet revolution, invisible yet seismic that shook the very foundation of reality. The classical world, with its comforting certainties, began to crumble at the edges as physicists probed ever deeper into the heart of matter. What they found was something wholly unexpected — a world not of fixed, deterministic laws but of probabilities, contradictions, and interconnectedness.

At the centre of this revolution lay the discovery of the quantum. It is a word that slips easily from the tongue, and yet it carries within it the weight of the universe. The quantum is the indivisible unit, the smallest quantity of energy or matter, the very building block from which all things arise. And yet, despite its diminutive size, the quantum holds within it secrets far greater than any that could be contained within the boundaries of classical physics.

One of the earliest and most unsettling of these secrets is that of wave-particle duality. In the quantum world, the fundamental constituents of reality — the electrons, photons, and other subatomic particles — are not solid objects, like marbles or grains of sand, as we might imagine. They are, instead, entities of dual nature, behaving sometimes as particles and at other times as waves. When we observe them closely, they appear to be discrete points localised in space and time. But when we step back, allowing them to move freely, they spread out like waves, flowing through space, their paths uncertain and diffuse.

The notion of a particle that is also a wave is an affront to common sense. How can something be both here and there at once? How can it move through space and yet remain undefined, its position and momentum smeared across a field of probabilities? These questions vexed the greatest minds of the twentieth century, forcing them to reconsider the very nature of reality itself. For in the quantum realm, it seems, particles do not occupy fixed positions in space and time until they are observed. Until then, they exist in a state of potentiality, a shimmering cloud of possibilities that collapses into a definite state only when measured.

This brings us to one of the most perplexing and

paradoxical features of quantum physics — the observer effect. In classical physics, the universe was thought to operate independently of the observer. Whether or not we chose to look, the world followed its preordained course, its laws immutable and indifferent to our gaze. But in the quantum world, this is no longer the case. The act of observation, it seems, alters the very nature of reality itself.

Consider, if you will, the famous double-slit experiment, a simple yet profoundly illuminating test of quantum behaviour. In this experiment, a beam of particles, such as photons or electrons, is fired at a barrier with two narrow slits. On the other side of the barrier is a screen, which records the pattern of particles that pass through the slits. When the experiment is run without any observation, the particles behave like waves, creating an interference pattern on the screen — a series of bright and dark bands as the waves from the two slits overlap and interfere with one another.

But when we observe the particles as they pass through the slits — when we attempt to measure which slit each particle travels through — something remarkable happens. The interference pattern disappears, and the particles behave as though they are solid objects, passing through one slit or the other but not both. The mere act of observation has changed their behaviour, forcing them to choose between being a wave or a particle.

This experiment, so simple in its design, strikes at the heart of the quantum mystery. It suggests that reality is not fixed but fluid, shaped by our interaction with it. The observer and the observed are not separate entities but are intertwined in a dance of mutual influence. In the quantum world, there is no objective reality, only a web of probabilities waiting to be collapsed into existence by the conscious mind.

And what of superposition? This strange phenomenon, perhaps the most bewildering of all quantum behaviours, defies all logic and reason. In the classical world, an object exists in one place at one time. A chair is a chair sitting in the corner of the room, occupying a single, well-defined position. But in the quantum world, particles can exist in multiple states simultaneously. An electron, for example, can be in two places at once, spinning in two directions at once until we observe it. Only then does it "choose" a single state.

This concept of superposition is famously illustrated by the thought experiment known as Schrödinger's Cat. Imagine a cat placed inside a sealed box, along with a radioactive atom, a Geiger counter, and a vial of poison. If the Geiger counter detects the decay of the atom, the poison is released, killing the cat. If no decay is detected, the cat remains alive. But until the box is opened and the system is observed, the cat exists in a state of superposition — it is both alive and dead at the same time, suspended between two contradictory realities.

While this thought experiment may seem absurd, it captures the unsettling nature of quantum mechanics. At the most fundamental level, the universe does not operate according to the binary logic of either/or. Instead, it embraces a multitude of possibilities, all coexisting in a state of flux until the conscious mind intervenes. It is a world where contradictions are not only possible but necessary, where reality itself is a shifting mosaic of probabilities and potentialities.

And yet, even this is not the final word on the strange nature of the quantum realm. In this world of uncertainties, we find not only superpositions and observer effects but also the eerie phenomenon of quantum entanglement. When two particles become entangled, they are linked in such a way that the state of one particle is instantly correlated with the state of the other, no matter how far apart they are. If one particle is measured and found to be spinning clockwise, the other, even if it is on the opposite side of the universe, will be found to be spinning counterclockwise.

This instantaneous connection between entangled particles defies our conventional understanding of space and time. In the classical world, information cannot travel faster than the speed of light, and yet entangled particles seem to communicate with one another instantly, as though distance is irrelevant. It is as if, on some deeper level, the particles are not separate at all but are merely different manifestations of a single, unified whole.

What, then, are we to make of this quantum world, where the laws of physics dissolve into paradox and mystery? Some would argue that it is simply a reflection of our incomplete understanding of the universe, a challenge for future generations of scientists to unravel. Others, however, see in quantum physics a profound truth about the nature of reality itself — that the universe is not a machine to be dissected and understood but a living, breathing entity, rich with mystery and wonder.

It is tempting, in the face of such complexity, to retreat into the safety of the known, to dismiss the quantum world as a mere curiosity, a playground for physicists and mathematicians. But to do so would be to turn away from the most profound insights into the nature of existence that humanity has ever glimpsed. For quantum physics, with all its paradoxes and contradictions, offers us a glimpse into the heart of reality, a reality that is far stranger and more wondrous than we could ever have imagined.

In the centuries to come, as we continue to probe the depths of the quantum realm, we may yet discover that the mysteries of the universe are not to be found in the equations of mathematics or the calculations of science but in the very act of being itself. The quantum world, with its entangled particles and superpositions, may not be a separate domain, hidden from our everyday experience, but the very essence of reality, waiting to be realized.

And so, as we stand on the threshold of this new frontier, we must learn to embrace the uncertainty, to let go of our need for fixed answers and definitive truths. For in the quantum realm, as in life, it is not certainty but possibility that reigns supreme. It is a world not of things but of potentials, where every moment is a gateway to new realities, new dimensions of existence waiting to unfold.

In this fluid dance of probabilities, we may come to realize that the boundaries between the observer and the observed, the self and the world, are not as rigid as they once seemed. Consciousness, that mysterious flame that burns within each of us, may be the key to unlocking the deepest secrets of the quantum realm. For in the act of observation, in the simple awareness of being, we shape the very fabric of reality, collapsing the infinite into the finite, and turning the possible into the real.

CHAPTER 3: THE NATURE OF REALITY – CLASSICAL VS. QUANTUM VIEWS

There is a certain allure to the universe that we believe we know, the universe composed of stable certainties, of solid ground beneath our feet and predictable patterns of stars wheeling above. For eons, humanity has clung to this vision of reality, one shaped by familiar rules, where time marches forward with relentless precision and objects remain steadfastly in their places until acted upon. It is a reality that has nourished our imaginations, a cosmos governed by laws as ancient as the stars themselves, laws that cradled the birth of suns and shaped the flight of birds.

But what is this reality, this edifice of apparent solidity

that we so confidently inhabit? Is it as permanent, as immutable, as we have long supposed? Or does it veil beneath its surface a deeper, more elusive truth, a truth that shimmers and shifts as we draw closer, evading the grasp of our reason like water slipping through our fingers?

In the classical view of the universe — the world that Isaac Newton and his intellectual descendants gave to us — reality is a structured and mechanistic thing. It is a grand machine, intricately wrought from particles of matter that move in accordance with laws that are fixed and immutable. The planets spin in their orbits, the apples fall from their branches, and the celestial and the earthly alike obey the same unyielding rules. There is no room for uncertainty in this world, no place for ambiguity. Everything that exists is measurable, tangible, calculable.

We must not begrudge the classical view its triumphs. For centuries, it illuminated the path of human progress, allowing us to harness the forces of nature, to navigate the seas, to build cities that scraped the skies. Newton's universe of order and causality was a gift, a vision of a cosmos in which every effect had a cause, and every action had a reaction. It was a world of predictable certainty, a universe where the future could be foretold with the same precision as the movements of a pendulum.

And yet, as we stand now at the threshold of the quantum world, we are forced to confront an uncomfortable question: is this classical reality the whole of existence, or is it but a shadow cast upon the walls of a deeper, more inscrutable cosmos? For as we delve into the subatomic depths, we find that the comforting certainties of the classical view begin to dissolve, giving way to a universe that is far more fluid, far more mysterious, than we had ever imagined.

The classical view tells us that the universe is composed of discrete particles, each with its defined position and velocity, each moving through space in a precise and orderly fashion. But in the quantum realm, this certainty vanishes like mist at dawn. Here, particles are not solid entities but wavering probabilities, existing in a multitude of states at once, flickering between possibilities like moths caught in the glow of a distant flame. It is a world where certainty gives way to potentiality, where the boundaries between being and non-being blur into a shimmering haze.

In the classical world, time flows like a river, moving inexorably from past to present to future, its current smooth and unbroken. It is the great lawgiver of the universe, the force that orders all things, shaping the rhythm of existence itself. But in the quantum realm, time becomes a far more elusive thing. The arrow of time, once thought to be unidirectional, may curve and bend, folding in upon itself in strange and unexpected ways. Particles may leap forward or backward in time, as if the past and the future are not fixed points but malleable constructs, shaped by the act of observation itself.

The implications of this are staggering, for they call into question the very nature of reality. If time is not a constant, if particles do not exist in defined states until they are observed, then what does it mean to say that something is real? Can we still speak of an objective reality, a reality that exists independently of our perception, or must we acknowledge that reality is something far more fluid, something shaped by the act of observation itself?

To grapple with these questions is to walk a tightrope between reason and madness, between what we have long accepted as true and what we now glimpse as possible. For in the quantum world, the observer is not a passive entity, standing apart from the universe, recording its movements with detached precision. The observer is a participant, an actor in the great drama of existence, whose very act of observation alters the course of events.

Consider the phenomenon of *quantum superposition*, that strange state in which particles exist in multiple places or states at once, as if they are caught between realities. In the classical world, a particle must occupy a single position; it must be here or there, never both. But in the quantum realm, particles defy this logic. They exist in a liminal state, suspended between possibilities, until the act of observation collapses them into a single reality.

It is as if the universe itself waits, poised on the brink of becoming, until the gaze of consciousness touches it, solidifying it into existence. Reality, then, is not a fixed thing, but a process, a continuous unfolding shaped by the interaction between the observer and the observed. The boundaries between subject and object, between self and world, begin to dissolve, revealing a deeper unity beneath the surface of things.

But this unity is not the harmonious order of the classical view. It is a unity born of paradox and uncertainty, a world where particles can be entangled across vast distances, where the actions of one particle can instantaneously affect another, no matter how far apart they may be. This phenomenon, known as *quantum entanglement*, challenges our very understanding of causality and locality, suggesting that the fabric of the universe is woven from threads that connect all things in ways that transcend space and time.

In this quantum universe, reality is not a static thing, but a dance, a ceaseless interplay of forces and probabilities, where each moment is both defined and undefined, where the act of observation crystallizes the infinite into the finite. It is a world where the future is not fixed, where the past is not immutable, where reality itself is in flux, shifting and shimmering like the surface of a lake caught in the light of a thousand suns.

The classical view, for all its elegance and precision, cannot contain this new reality. It is a world too rigid, too constrained by the laws of cause and effect, to accommodate the fluidity of the quantum realm. And yet, we cannot entirely discard the classical view, for it remains a useful approximation of the macroscopic world, the world we experience in our everyday lives. The classical and the quantum, then, are not opposites, but complementary perspectives, each revealing a different facet of the cosmos.

To navigate between these two worlds is to embrace a new kind of knowing, one that is not bound by the limitations of classical logic but is open to the possibilities of paradox. In the quantum world, opposites can coexist, contradictions can be true, and the line between reality and illusion becomes thin and porous. It is a world that demands of us not certainty, but humility, an acknowledgment that reality is far stranger, far more mysterious, than we could ever have imagined.

And yet, for all its strangeness, the quantum world is not alien to us. For within its paradoxes, within its shimmering probabilities, we find echoes of our own experience, our own consciousness. Just as particles exist in multiple states at once, so too do we, as conscious beings, inhabit a world of multiple possibilities, a world shaped by our perceptions, our choices, our actions. Just as time is fluid in the quantum realm, so too is time fluid in the realm of the mind, where memories of the past and visions of the future can merge and mingle in the present moment.

In this sense, the quantum world is not a departure from reality, but a deeper exploration of it. It is a reminder that reality is not a fixed and static thing, but a dynamic and evolving process, one that we are an integral part of. The classical view gave us a universe of order and certainty, a universe in which we were passive observers, watching the great machine of existence unfold according to its own predetermined laws. But the quantum view invites us to see ourselves as participants in this unfolding, as co-creators of reality itself.

It is a daunting invitation, for it requires us to let go of the comforting certainties that have long sustained us. It requires us to embrace uncertainty, to accept that reality is not something that is given to us, but something that we must actively engage with, something that we must help to create. But in this uncertainty, there is also great freedom, for it opens the door to new possibilities, new ways of being, new ways of knowing.

In the classical world, the future was a fixed and predetermined thing, a consequence of the laws of physics unfolding inexorably through time. But in the quantum world, the future is not fixed; it is a landscape of possibilities, a field of potentialities waiting to be realized. Each moment is a choice, a collapse of the wave function, a crystallization of possibility into reality. And in this collapse, we find not only the birth of the universe, but the birth of consciousness itself.

For consciousness, like reality, is not a static thing. It is a process, a continuous unfolding, shaped by our interactions with the world around us. In the quantum view, the boundaries between self and world, between mind and matter, are not as rigid as we once believed. Consciousness is not something that happens *within* us, but something that happens *between* us and the world, something that arises in the interplay between observer and observed, between self and other.

And so, as we contemplate the nature of reality, we must also contemplate the nature of consciousness. For the two are inextricably linked, bound together in the dance of existence. The quantum world challenges us to rethink not only our understanding of the universe, but our understanding of ourselves. It invites us to see that we are not separate from the world, not passive spectators in the theater of existence, but active participants, co-creators of the reality we inhabit.

In this new vision, reality is not something that we discover, but something that we help to create. It is a living, breathing thing, a dynamic and evolving process, shaped by the interplay of forces, particles, and probabilities. And at the heart of this process lies consciousness, the mysterious flame that lights the path of existence, the spark that brings the universe into being.

As we stand on the cusp of this new understanding, we are faced with a choice. We can cling to the certainties of the classical world, to the comforting illusion of a universe governed by fixed laws and immutable truths. Or we can embrace the fluidity of the quantum realm, with all its uncertainties and paradoxes, and in doing so, open ourselves to new possibilities, new ways of being, new ways of knowing.

It is a choice that each of us must make, not once, but again and again, in every moment of our lives. For reality is not something that happens to us; it is something that we create, moment by moment, choice by choice, breath by breath. And in this act of creation, we may come to realize that the nature of reality is not something to be discovered, but something to be lived.

CHAPTER 4: THE OBSERVER EFFECT AND CONSCIOUSNESS

To gaze into the depths of the cosmos is to confront a paradox: the universe, vast and indifferent, seems to operate with a kind of detached perfection, yet it changes, subtly and profoundly, when observed by the mind of a conscious being. This is the essence of the observer effect, a whispering riddle in the heart of quantum mechanics that draws us into a labyrinth of thought, where the boundaries between reality and perception blur, and the very act of observation becomes an act of creation.

The ancient sages, in their quiet contemplation, long suspected that the mind and the world were more intimately entwined than the senses would suggest. The philosophers of old understood that there was something ineffable about perception, something that hinted at a deeper connection between the observer and the observed. But it was only with the advent of quantum physics that this suspicion was transformed into something far more disquieting, something that shattered the certainties of classical thought and ushered in a new era of inquiry.

The observer effect, in its most distilled form, reveals to us a universe that is not passive, not a stage upon which the drama of existence plays out independently of those who witness it. Instead, it is a universe that is participatory, responsive, mutable — a universe that becomes only as it is seen. It is as if reality itself lies in wait, suspended in a state of potentiality, until the gaze of the observer calls it forth from the shadows of possibility into the light of existence.

What does it mean, then, to observe? To what extent does the mere act of perception alter the fabric of the world? And, perhaps most unsettling of all, what does this say about the nature of reality itself? If the universe depends, in some profound way, upon the consciousness that perceives it, then where do we draw the line between mind and matter, between the self and the world? These are not questions to be answered easily, for they touch upon the very core of what it means to exist, to be conscious, to be alive.

In the classical view, reality was thought to be objective, existing independently of the observer. The apple falls from the tree whether we are there to witness its descent or not; the moon circles the earth whether or not we cast our eyes upon it. The universe, in this view, is a great and intricate machine, set in motion by forces that we may observe, but which operate irrespective of our presence. But in the quantum world, this comforting distinction begins to break down. Here, the presence of the observer is not incidental; it is fundamental.

One of the most striking illustrations of the observer effect is found in the double-slit experiment, an elegant and perplexing demonstration of quantum behaviour. In this experiment, particles such as photons or electrons are fired at a barrier with two narrow slits, and a screen behind the barrier records the pattern formed by the particles that pass through. When the particles are unobserved, they behave as waves, creating an interference pattern on the screen — bright and dark bands that suggest the particles passed through both slits simultaneously, as waves would. But when the particles are observed, when a measuring device is placed at the slits to detect which path the particles take, they behave as particles, passing through one slit or the other, and the interference pattern disappears.

This shift — this change in behaviour based on observation — is more than a mere curiosity. It speaks to something deeply unsettling about the nature of reality itself. How can particles behave as both waves and particles, depending on whether or not they are observed? How can the act of measuring their position collapse their wave-like potentiality into the definite state of a particle? The observer, it seems, is not merely a passive witness to events, but an active participant in the unfolding of reality. The act of observation is an act of determination, a moment in which possibilities are collapsed into certainties, and potential is transformed into actuality.

To observe, then, is to create. In the quantum world, reality does not unfold independently of the observer; it is shaped, moulded, brought into being by the act of observation itself. The universe, far from being a fixed and unchanging thing, is in constant dialogue with consciousness, responding to the gaze of the observer with a kind of fluidity and malleability that defies all our previous notions of objectivity.

But what is this gaze? What is it that transforms mere possibility into reality? It is not simply the act of measurement, for even the most sophisticated instruments are nothing without the conscious mind that interprets their results. No, it is consciousness itself that stands at the heart of the observer effect, that mysterious and intangible force that gives rise to the world we perceive. Consciousness, the luminous essence that animates our thoughts and feelings, is not a passive bystander in the drama of existence; it is an actor, a creator, a force that shapes the very fabric of reality.

In this sense, the observer effect is not merely a phenomenon of quantum physics; it is a window into the nature of consciousness itself. For just as particles exist in a state of superposition, suspended between multiple possible states until they are observed, so too does consciousness exist in a state of potentiality, a field of infinite possibility that is collapsed into definite thoughts, feelings, and experiences by the act of attention. Consciousness is not a thing, not an object to be dissected and analysed; it is a process, a becoming, an unfolding that shapes the world as it shapes itself.

To be conscious is to observe, and to observe is to create.

But this creation is not a one-way street; it is a dance, a dialogue between the observer and the observed, between the self and the world. The observer effect reveals to us that reality is not something that exists independently of us, but something that we participate in, something that we bring into being through the act of observation.

This raises profound questions about the nature of free will and agency. If reality is shaped by consciousness, then to what extent are we responsible for the world we inhabit? Do we create reality with each thought, each perception, each moment of attention? Or is there a deeper reality that lies beyond the reach of our consciousness, a reality that exists independently of our gaze? These are questions that quantum physics cannot answer, for they belong to the realm of metaphysics, to the ancient and timeless inquiry into the nature of existence itself.

And yet, even in the face of these unanswered questions, there is something deeply empowering about the observer effect, for it suggests that we are not mere cogs in the great machine of the universe, not passive spectators in a predetermined drama. We are creators, participants in the unfolding of reality, capable of shaping the world through the act of observation. Our consciousness is not a fleeting spark in the void; it is a force that shapes the very fabric of existence.

But if we are to accept this, if we are to embrace the idea that consciousness plays a fundamental role in shaping reality, then we must also accept the responsibility that comes with it. For to observe is not a neutral act; it is an act of creation, an act of determination that brings possibilities into being. We cannot separate ourselves from the world we observe, for in the act of observation, we are inextricably bound to it.

This realization calls into question the very nature of subjectivity and objectivity. In the classical view, the observer and the observed were distinct, separate entities. The observer was a neutral, detached presence, capable of recording the events of the world without influencing them. But the quantum world reveals to us a deeper truth: the observer is not separate from the observed; the two are bound together in a dynamic interplay that shapes the very fabric of reality.

The boundary between subject and object, between self and world, begins to dissolve. In the quantum world, there is no clear distinction between the observer and the observed, for the two are inextricably intertwined. Consciousness is not a mere reflection of reality; it is a force that shapes it, moulds it, brings it into being.

This dissolution of boundaries is not merely a scientific observation; it is a profound philosophical insight, one that echoes the wisdom of the mystics and sages of old. For millennia, the spiritual traditions of the world have taught that the self and the world are not separate, that consciousness and reality are one. The observer effect, in its own way, affirms this ancient wisdom, revealing to us a universe in which consciousness and reality are not distinct but are part of the same unfolding process.

To live in such a universe is to embrace a new kind of responsibility, for it is not only the physical world that is shaped by our observations, but the moral and spiritual world as well. If consciousness plays a role in shaping reality, then our thoughts, our perceptions, our attention, carry with them a weight that extends far beyond the confines of the self. To observe is to shape, and to shape is to create.

And so, the observer effect calls us to a new kind of awareness, an awareness not only of the world around us but of the role we play in its creation. We are not passive spectators in the drama of existence; we are active participants, creators of reality through the act of observation. Our consciousness is not a fleeting spark in the vastness of the cosmos; it is a force that shapes the very fabric of the universe.

In the silence that follows this realization, we are left with a choice. We can retreat into the comforting certainties of the classical world, where reality exists independently of our gaze, or we can embrace the fluidity and potentiality of the quantum world, where consciousness and reality are intertwined in a dynamic dance of creation. The choice is not an easy one, for it requires us to relinquish the illusion of certainty, to accept that the world we inhabit is shaped, in part, by the act of observation itself.

But in this acceptance, there is also freedom, for it opens the door to new possibilities, new ways of being, new ways of knowing. The observer effect invites us to see the world not as a fixed and unchanging thing, but as a process, a becoming, an unfolding that is shaped by our consciousness. It invites us to step into the role of creators, to embrace the responsibility that comes with the act of observation, and to recognize that the world we perceive is, in part, a reflection of the consciousness that perceives it.

In the quiet moments of contemplation, we may come to realise that the universe is not a distant and indifferent thing, not a machine that operates independently of us, but a living, breathing entity, a universe that responds to the gaze of consciousness, a universe that becomes as it is seen.

CHAPTER 5: THE HOLOGRAPHIC UNIVERSE AND THE NATURE OF REALITY

There are whispers in the deep silence of existence, murmurs that hum beneath the fabric of reality, intimating that the universe itself is but a reflection, an illusion of light and shadow cast upon an unfathomable screen. To live in such a cosmos is to dwell within a grand illusion, a hall of mirrors where the flickering image of solidity is not what it seems, where the boundaries between the physical and the ethereal dissolve into a shimmering haze of suggestion. This is the world of the holographic universe, a theory so profound, so strange, that it bends our understanding of existence itself, leaving us suspended between wonder and uncertainty.

To speak of the universe as holographic is to unravel the very notion of reality, to step beyond the limits of perception and into a domain where the physical is but a mask for something deeper, something ineffable. It is an invitation to contemplate the possibility that what we perceive as space and time, as matter and form, are but projections of a deeper, unseen reality, a reality that lies beyond the reach of our senses and yet shapes the very nature of existence.

In the traditional view, the universe is a vast expanse, a seemingly infinite domain of matter and energy governed by immutable laws. It is a world of separation, where objects exist independently of one another, occupying distinct locations in space and time. We navigate this world with the certainty that what we see is what is — that the stars shining above are as real as the ground beneath our feet, that the objects we touch are solid and substantial.

But the holographic principle offers a different vision, one in which the entire universe is a projection, a shadow cast from a deeper dimension. It suggests that the reality we experience — the three-dimensional world of matter and form — is a mere surface, a holographic image encoded on the boundary of a higher-dimensional space. Just as a hologram contains the complete information of a three-dimensional object on a two-dimensional surface, so too does the universe contain within it the imprint of a higher reality, a reality that transcends the limitations of space and time.

This notion, at first glance, may seem like a fantasy, a

dream spun from the imagination of poets and mystics. But it is rooted in the deepest inquiries of modern physics, in the strange and unsettling discoveries of quantum mechanics and string theory. It is a theory born not from whimsy but from the very heart of scientific rigor, a theory that challenges our most fundamental assumptions about the nature of reality.

To understand the holographic principle, we must first consider the nature of information. In the classical view, information is something that can be stored and transmitted, something that exists as a discrete quantity. A book contains information in the form of words and sentences; a computer stores information as bits of data. But in the quantum world, information takes on a far more subtle and profound role. It becomes the very fabric of reality itself, the underlying code that shapes the universe.

In the holographic model, the universe is not composed of solid objects but of information. The particles that make up matter — the electrons, quarks, and photons that dance through the quantum field — are not independent entities but are, instead, expressions of a deeper informational structure. This structure is encoded on the boundary of the universe, on a surface that contains all the information necessary to describe the entire cosmos.

It is as if the universe were a vast cosmic hologram, with every particle, every star, every galaxy encoded on its surface. The three-dimensional world we experience is but the unfolding of this information, a projection that gives the illusion of depth and form. And just as a hologram contains the complete image of an object in every part, so too does the universe contain the whole within each of its parts. Every particle, every atom, carries within it the imprint of the entire cosmos, reflecting the whole in its smallest fragment.

This idea, that the universe is holographic, is not merely a theoretical abstraction. It has profound implications for our understanding of space and time, of matter and energy, of consciousness itself. It suggests that the boundaries we perceive — the separations between self and other, between mind and world — are illusions, that the reality we experience is but a reflection of a deeper unity that transcends these divisions.

In this holographic universe, the distinction between the part and the whole dissolves. Just as each fragment of a hologram contains the complete image of the whole, so too does each moment, each particle, each thought contain the entirety of existence within it. The universe is not a collection of separate objects, but a seamless, interconnected whole, a cosmic web in which every point is linked to every other point. To perceive the universe in this way is to see the world as a single, indivisible reality, a reality in which the distinctions between past and future, between here and there, between self and world, are but illusions.

This dissolution of boundaries extends beyond the physical realm into the realm of consciousness. For if the universe is holographic, then consciousness itself is part of this hologram, a projection of the deeper informational structure that underlies reality. The mind is not an isolated entity, separate from the world it perceives; it is a part of the holographic whole, a reflection of the same cosmic information that shapes the physical universe.

In this view, consciousness is not confined to the brain, not limited to the boundaries of the self. It is a field, an unfolding of information that spans the entire universe. Just as each particle contains the imprint of the whole, so too does each conscious mind contain the imprint of the cosmos, reflecting the totality of existence within its perceptions and experiences. The mind is not a passive observer of reality but an active participant, shaping and being shaped by the holographic field in which it is embedded.

But what does it mean to live in such a universe? What does it mean to inhabit a world that is not solid and fixed but fluid and malleable, a world in which reality is a projection, an illusion shaped by the deeper currents of information that flow through the cosmos?

To live in a holographic universe is to embrace a new way of seeing, a way of perceiving the world not as a collection of separate objects but as an interconnected whole. It is to recognise that the boundaries we perceive are not absolute, that the separations between self and other, between mind and matter, are not fixed but fluid. In this universe, the distinction between the observer and the observed dissolves, revealing a deeper unity that underlies all of existence.

In the classical view of the universe, we are separate from the world, standing apart from it as observers, recording its events without influencing them. But in the holographic universe, we are part of the very fabric of reality, woven into the cosmic web of information that shapes the universe. Our thoughts, our perceptions, and our consciousness are not separate from the world; they are part of the same holographic field that gives rise to matter and energy.

This realisation has profound implications for how we understand reality, for it suggests that the world we experience is not something that exists independently of us but something that we help to create. The universe is not a fixed, objective reality that unfolds according to immutable laws; it is a dynamic, evolving process shaped by the interactions between the observer and the observed, between consciousness and matter.

In this holographic universe, reality is not a thing to be grasped, not a solid object to be dissected and analysed. It is a process, a becoming, a continuous unfolding of information that gives rise to the world we perceive. And just as a hologram contains the complete image of the whole in every part, so too does the universe contain within it the potential for infinite realities and infinite possibilities.

But this potential is not passive; it is shaped by the conscious mind, by the act of observation and perception. In the holographic universe, consciousness plays a fundamental role in shaping reality, in collapsing the infinite potential of the holographic field into the specific, finite realities we experience. The mind is not a bystander in the unfolding of the universe; it is a creator, a force that shapes the world through the act of perception.

This idea that consciousness plays a role in shaping reality echoes the insights of quantum mechanics, where the act of observation collapses the wave function, bringing potentiality into actuality. In the holographic universe, this process is extended to the cosmos itself, where consciousness and information are intertwined in the creation of reality.

But if reality is a projection, a holographic image encoded on the boundary of the universe, then what lies beyond this projection? What is the source of this holographic field, the deeper reality that gives rise to the world we experience?

This question takes us beyond the realm of physics and into the domain of metaphysics, into the ancient inquiry into the nature of existence itself. The holographic universe suggests that there is a deeper reality, a higher-dimensional space from which the universe as we know it is projected. This deeper reality is not bound by the limitations of space and time, not confined to the physical laws that govern the universe. It is a realm of pure information, a cosmic sea of potentiality that gives rise to the world of form and matter.

In this deeper reality, the distinctions between subject and object, between mind and matter, between self and world dissolve completely. There is no separation, no division between the observer and the observed. There is only the infinite unfolding of information, the continuous creation of reality through the interaction of consciousness and the holographic field.

To contemplate this deeper reality is to confront the limits of thought, for it lies beyond the reach of language and concept, beyond the boundaries of perception and experience. It is a reality that cannot be grasped by the mind, for the mind itself is a part of the holographic projection, a reflection of the deeper informational structure that shapes the universe.

And yet, even though this deeper reality lies beyond our grasp, it is not beyond our experience. For just as the holographic image contains the whole within each of its parts, so too does the universe contain within it the imprint of the deeper reality that gives rise to it. In the silence of contemplation, in the stillness of the mind, we may catch a glimpse of this deeper reality, a flicker of the infinite that lies beyond the boundaries of perception.

To live in a holographic universe is to embrace the mystery of existence, to recognise that the world we perceive is but a reflection, an illusion shaped by the deeper currents of information that flow through the cosmos. It is to understand that reality is not a fixed and solid thing but a fluid and dynamic process, a continuous unfolding of potentiality into actuality.

In this universe, the mind is not separate from the world but is an integral part of the holographic field, a participant in the creation of reality. Our thoughts, our perceptions, and our consciousness shape the world, collapsing the infinite potential of the holographic field into the specific realities we experience.

And in this realization, we may find a deeper truth, a truth that transcends the boundaries of space and time, a truth that lies at the heart of the holographic universe: that reality is not something that happens to us, but something that we help to create. We are not passive observers in the drama of existence, but active participants, co-creators of the world we inhabit. And in this act of creation, we may come to understand that the universe itself reflects consciousness, a holographic projection of the infinite potential that lies within us all.

CHAPTER 6: QUANTUM SOUL – A METAPHYSICAL PERSPECTIVE

In the silent expanse where the cosmos breathes and the stars pulse with an ancient rhythm, there exists a secret woven not in matter but in the invisible threads of consciousness. The notion of a soul, that flicker of divine light nestled within each living being, has been the subject of endless musings and speculations throughout the ages. Philosophers, mystics, and sages have grappled with its essence, each seeking to unveil the unfathomable depth of this elusive force. But what if the soul is not confined to the mystical or the spiritual alone? What if, embedded within the quantum fabric of existence, we find traces of the quantum soul, a force that transcends both the physical and the metaphysical, bridging the chasm between the realms of thought and the energies that shape reality itself?

To contemplate the quantum soul is to step into a liminal space, a twilight of reason and imagination where the familiar boundaries of physics dissolve into a fluid dance of potential. This is a domain where the soul does not merely inhabit the body nor linger as a ghost in the machine but where it intertwines with the very essence of quantum phenomena — with superposition, entanglement, and non-locality. Here, the soul, like the particles of the quantum world, is both localised and diffused, both here and elsewhere, a flickering light that transcends the confines of space and time.

There is, within this concept, the suggestion that the soul, far from being an ephemeral and separate entity, is woven into the fabric of the quantum universe. It is not merely a passenger in the vessel of the body but a participant in the ceaseless dance of energy and matter. As the physicists explore deeper into the heart of quantum mechanics, they find that the lines between what is and what could be are not as distinct as they once seemed. The soul, perhaps, exists not in defiance of the physical universe, but as an integral part of its most fundamental workings.

The classical world, with its rigid distinctions between the material and the immaterial, offered little space for the soul. It painted a picture of the universe as a great machine, where the mind was a byproduct of the brain, and the soul was either dismissed as superstition or relegated to the realms of theology. But in the quantum world, where particles can exist in multiple states at once, where reality itself seems to be shaped by the observer, there is room for a different understanding. Here, the soul is not confined to the shadows of metaphysics but steps boldly into the light of scientific inquiry.

Consider for a moment the concept of superposition, that strange and wondrous phenomenon in which particles exist in multiple states simultaneously until they are observed, until the gaze of consciousness collapses them into a single reality. In this light, the soul, too, may be seen as existing in a state of superposition, as being both here and elsewhere, as existing both within the body and beyond it. It is not a fixed and finite entity but a field of potentiality, a shimmering presence that transcends the boundaries of the physical and the metaphysical.

What if the soul, like the particles of the quantum world, is not bound by the constraints of space and time? What if it exists not in one place, but in many, its essence diffused throughout the cosmos, connected to every other soul, to every other particle in the universe through the invisible threads of quantum entanglement? In this vision, the soul is not an isolated spark of divinity, but part of a vast, interconnected web of consciousness, a web that stretches across the cosmos, linking all beings in a shared field of awareness.

The concept of entanglement, too, offers a tantalizing metaphor for the soul. In the quantum world, entangled particles, once they have interacted, remain connected no matter how far apart they may be. The state of one particle instantaneously affects the state of the other, even if they are separated by vast distances. This phenomenon, which defies the classical understanding of space and time, suggests that the universe is far more interconnected than we have ever imagined. And if particles can be entangled, then why not souls? What if, at the moment of birth, or perhaps even before, our souls become entangled with the souls of others, with the universe itself, forming an intricate web of connections that transcends the limitations of space and time?

In this web of entanglement, the soul is not a solitary traveller but a part of a greater whole, a node in the cosmic network of consciousness. It is through these connections that we experience empathy, love, and compassion, for the boundaries between self and other, between one soul and another, are not as fixed as they appear. The quantum soul is not confined to the individual but is part of a collective, a shared field of awareness that links all beings across the vastness of the cosmos.

This vision of the soul as an interconnected, quantum entity offers a profound challenge to the materialist view of consciousness, which holds that the mind is nothing more than the product of the brain, a fleeting phenomenon that ceases to exist at death. But if the soul is entangled with the quantum universe, if it is woven into the fabric of reality itself, then it cannot be so easily dismissed. It is not a product of the brain but a fundamental aspect of the cosmos, a force that transcends the physical and the temporal.

The soul, in this sense, may be seen as a kind of field, a field of consciousness that permeates the universe, interacting with the quantum field, shaping and being shaped by it. It is through this interaction that we experience the world that we perceive the richness and complexity of existence. The mind, then, is not an isolated phenomenon but a part of this greater field, a receiver and transmitter of consciousness that is connected to the quantum soul.

But what, then, is the nature of this consciousness? What is it that animates the quantum soul that gives it its vitality and its power? Here, we step into the deepest mysteries of existence, mysteries that have eluded the grasp of both science and philosophy for millennia. Consciousness, like the soul, is a force that transcends the physical, a force that cannot be reduced to the firing of neurons or the chemical reactions of the brain. It is a force that is woven into the very fabric of the universe, a force that shapes reality even as it is shaped by it.

In the quantum world, consciousness plays a fundamental role in shaping reality. It is the act of observation, the act of awareness, that collapses the wave function that brings potentiality into actuality. The universe, it seems, is not a fixed and static thing but a fluid and dynamic process, a process that is shaped by consciousness. And if consciousness plays such a role in shaping the physical world, then what role does it play in shaping the soul?

The quantum soul, in this sense, is not a passive entity but an active force, a force that shapes reality through the act of consciousness. It is through the soul that we interact with the quantum field that we collapse the potentialities of the universe into the specific realities we experience. The soul, then, is not a mere bystander in the drama of existence but a co-creator, a participant in the unfolding of reality.

This vision of the soul as an active, creative force is not entirely new. It echoes the insights of the mystics and the sages, who have long taught that the soul reflects the divine, a spark of the creative force that shapes the cosmos. But the quantum perspective offers a new lens through which to view this ancient wisdom, a lens that reveals the soul not as something separate from the physical world but as an integral part of it, as a force that is woven into the very fabric of reality.

In this light, the quantum soul is not confined to the individual, not limited to the self. It is part of a greater whole, a collective consciousness that spans the cosmos, a field of awareness that links all beings in a shared experience of existence. This collective consciousness is not static but dynamic, constantly evolving, interacting with the quantum field to shape the reality we experience.

But if the soul is part of this collective consciousness, then what happens to it after death? What becomes of the quantum soul when the body ceases to function, when the brain no longer fires its neurons when the self dissolves into the void? Here, again, the quantum perspective offers a new way of understanding the mystery of death and the afterlife.

In the classical view, death is the end, the cessation of consciousness, and the dissolution of the self. But in the quantum world, where particles exist in superposition, where time is fluid, and the boundaries between past, present, and future are not as fixed as they seem, death may not be so final. The quantum soul, entangled with the cosmos, may persist beyond the death of the body, existing in a state of potentiality, a field of consciousness that transcends the limitations of space and time.

This does not mean that the self, as we know it, continues after death. The self, with its thoughts and memories, desires and fears, is a product of the brain, a construct of the mind. But the quantum soul, the deeper field of consciousness that underlies the self, may continue to exist, interacting with the quantum field, shaping and being shaped by the universe in ways that we cannot fully understand.

In this vision of the afterlife, the soul does not ascend to a higher realm, nor does it dissolve into nothingness. It persists, not as a static entity but as a dynamic field of consciousness, a force that continues to interact with the quantum field that continues to shape and be shaped by the universe. The soul, in this sense, is not bound by the limitations of the physical world but exists in a state of superposition, a state of potentiality that transcends the boundaries of life and death.

To contemplate the quantum soul is to confront the deepest mysteries of existence, to grapple with questions that have eluded the grasp of both science and philosophy for millennia. It is to step beyond the limitations of the classical world, beyond the rigid distinctions between the physical and the metaphysical, and into a realm where the soul is not separate from the universe but is woven into its very fabric, a force that shapes reality even as it is shaped by it.

In this quantum world, the soul is not a passive entity, not a ghost in the machine, but an active force, a creative power that interacts with the quantum field to shape the reality we experience. It is a field of consciousness that transcends the boundaries of the self, a collective awareness that links all beings in a shared experience of existence. And it is through this collective consciousness that we experience empathy, love, and compassion, for the boundaries between self and others are not as fixed as they appear.

In the stillness of contemplation, we may catch a glimpse of this deeper reality, a flicker of the infinite that lies beyond the boundaries of perception. We may come to understand that the quantum soul is not something that happens to us but something that we help to create, something that we participate in, moment by moment, thought by thought, breath by breath. Through this act of creation, we may come to understand that the soul is not a static entity but a dynamic process, a field of consciousness that is constantly evolving, constantly interacting with the quantum field to shape the reality we experience.

CHAPTER 7: ENTANGLEMENT AND THE INTERCONNECTEDNESS OF SOULS

There exists a hidden thread, invisible to the eye, intangible to the touch, that weaves through the fabric of existence, binding all things in a delicate web of relation. It is the unseen bond that links the stars to the sea, the bird to the wind, and the heart of one to the heart of another. This thread, ethereal and mysterious, has long eluded the grasp of reason, yet its presence is undeniable, whispered of by poets, dreamed of by mystics, and now, touched upon by the delicate fingers of quantum theory. This is the thread of entanglement, a phenomenon that transcends the physical world and reaches deep into the core of being, suggesting that, at its most fundamental level, reality is not composed of discrete entities but of profound and inseparable connections.

Entanglement, in the language of the quantum world, speaks of particles that, having once interacted, remain intertwined regardless of distance. They become mirror reflections of each other, their states forever synchronised so that the actions of one reverberate instantaneously upon the other, even across the vast chasms of space. It is as though the universe, in its quiet wisdom, has whispered into the very heart of reality that separation is but an illusion, that nothing exists in isolation, and that every atom, every breath, every soul, is connected in ways that defy our deepest understanding.

To contemplate the soul in light of entanglement is to glimpse a reality far grander than the self-contained individuality to which we so often cling. The soul, in this view, is not a solitary flame, flickering alone in the void, but part of a vast and radiant field of interconnected consciousness. It is entangled with the universe itself, bound by invisible threads to every other soul, to every particle, to the very cosmos in which it resides. In this web of entanglement, the boundaries between self and other dissolve, and we find ourselves not as separate beings but as expressions of a deeper unity, a unity that transcends the limitations of space, time, and matter.

In the ancient days, when the mystics first sought to understand the nature of the soul, they often spoke of a divine oneness, a state of being in which all things were connected, where the individual self melted into the greater whole. This idea of oneness, of interconnectedness, was not merely a spiritual ideal but a reflection of a deeper truth about the nature of reality, a truth that has now been echoed in the discoveries of quantum physics. In the quantum world, the very particles that form the building blocks of matter are not separate, isolated entities. They are entangled, linked in ways that transcend the physical, creating a web of relations that stretches across the entire universe.

What does this mean for the soul? If particles can be entangled, then why not consciousness itself? What if, at the deepest level of existence, our souls are not separate, independent entities but are instead entangled with one another, connected through the invisible threads of the quantum field? What if every thought, every emotion, every moment of awareness sends ripples through this field, affecting not only ourselves but every other soul to which we are connected?

This is not merely a metaphorical entanglement but a literal one, a connection that exists at the very heart of reality. It suggests that the soul is not confined to the body, not limited to the individual self. It reaches beyond the boundaries of the physical, extending into the quantum field, where it is linked to every other soul, to the very cosmos itself. In this vision, the soul is not a static, isolated spark of consciousness but a dynamic, living force that is constantly interacting with the universe, constantly shaping and being shaped by the interconnected web of existence.

To understand this is to recognise that we are not alone, not separate from the world around us. Every thought, every action, every breath is part of a larger whole, part of the vast network of consciousness that binds all things together. This interconnectedness is not a distant or abstract concept but a lived reality, a truth that is woven into the very fabric of existence. We are entangled with one another, not just in the metaphorical sense of shared experience or common humanity, but in the most literal, physical sense. Our souls, like the particles of the quantum world, are linked, bound together by the invisible forces that shape reality.

In the classical view of the universe, the self is seen as an isolated entity, a being that exists apart from the world, interacting with it but remaining fundamentally separate. But the quantum view, with its emphasis on entanglement and interconnectedness, offers a different perspective. It suggests that the self is not an isolated entity but a part of a larger whole, a node in the cosmic web of consciousness. The boundaries between self and other, between mind and matter, between soul and universe, are not fixed but fluid, shifting and changing with every moment of awareness.

This fluidity is mirrored in the phenomenon of entanglement itself. In the quantum world, particles that have become entangled are not bound by the limitations of space and time. They exist in a state of non-locality, meaning that their connection is instantaneous, regardless of distance. This suggests that the universe is not a collection of separate objects but a unified whole, a web of interconnected relations that transcends the physical boundaries of space and time.

What does this mean for the nature of the soul? It suggests that the soul, too, is not bound by the limitations of the physical world. It exists in a state of non-locality, connected to every other soul, to every particle, to the entire cosmos. This non-locality is not a distant or abstract concept but a lived reality, a truth that is woven into the very fabric of existence. Our souls, like the particles of the quantum world, are linked, bound together by the invisible forces that shape reality.

This idea of non-locality challenges the classical view of the soul as a separate, isolated entity. It suggests that the soul is not confined to the body, not limited to the self. It reaches beyond the boundaries of the physical, extending into the quantum field, where it is linked to every other soul, to the very cosmos itself. In this vision, the soul is not a static, isolated spark of consciousness but a dynamic, living force that is constantly interacting with the universe, constantly shaping and being shaped by the interconnected web of existence.

If the soul is entangled with the cosmos, then it is not subject to the limitations of space and time. It exists in a state of potentiality, a field of infinite possibility that is shaped by the act of consciousness. The soul is not bound by the constraints of the physical world but is free to move, to change, to evolve. It is a force that is constantly interacting with the quantum field, shaping reality through the act of awareness.

In this vision of the soul, consciousness is not a passive phenomenon but an active force, a creative power that shapes the universe. The soul, in this sense, is not a mere observer of reality but a participant in its creation. It is through the act of awareness that the soul interacts with the quantum field, collapsing the infinite potential of the universe into the specific realities we experience.

But if the soul is entangled with the universe, if it is part of the interconnected web of existence, then what happens when it encounters another soul? In the classical view, the self and the other are seen as separate, distinct entities, interacting with one another but remaining fundamentally isolated. But in the quantum world, where entanglement and non-locality reign, the boundaries between self and other dissolve.

When two souls meet, their entanglement is not merely a connection of bodies or minds but a connection of consciousness, a link that transcends the physical and reaches into the quantum field. The encounter between two souls is not a simple exchange of words or thoughts but a profound interaction that resonates through the web of existence, sending ripples through the quantum field, affecting not only those involved but every other soul to which they are connected.

This entanglement of souls is not limited to moments of direct interaction. It extends beyond the boundaries of space and time, linking us to others in ways that we cannot fully understand. The people we love, the people we encounter, even those we may never meet, are part of this vast web of entanglement. Their thoughts, their emotions, and their very being are connected to ours through the invisible threads of the quantum field.

This interconnectedness has profound implications for how we live, how we love, and how we understand ourselves. It suggests that the boundaries we place between self and other, between mind and matter, between soul and world, are not as fixed as they seem. We are not separate beings, isolated in our own experiences, but part of a larger whole, part of the vast and intricate web of consciousness that binds the universe together.

In this vision, empathy, compassion, and love are not merely emotions but reflections of a deeper truth about the nature of reality. They are the expressions of the entanglement of souls, the recognition that we are all connected, that our actions, our thoughts, and our very being affect not only ourselves but every other soul to which we are linked. To love, in this sense, is to acknowledge the entanglement of souls, to recognise that the boundaries between self and other are illusory and that we are all part of the same cosmic web of consciousness.

The entanglement of souls is not a static or passive phenomenon but a dynamic, evolving process. Just as particles in the quantum world are constantly interacting, constantly influencing one another, so too are our souls constantly interacting with the universe, shaping and being shaped by the web of existence. This entanglement is not something that happens to us; it is something we participate in, something we help to create through the act of consciousness.

In the quiet moments of reflection, we may catch a glimpse of this deeper reality, a flicker of the interconnectedness that lies beneath the surface of existence. We may come to understand that the soul is not a solitary traveller but part of a greater whole, a node in the cosmic web of consciousness. And in this understanding, we may find a new way of being, a new way of loving, a new way of understanding ourselves and the world around us.

To live in the awareness of the entanglement of souls is to recognise that we are not separate, not isolated, not alone. We are part of a larger whole, part of the vast and intricate web of consciousness that binds the universe together. Our souls are not confined to the body, not limited to the self. They reach beyond the boundaries of the physical, extending into the quantum field, where they are linked to every other soul, to the entire cosmos.

In this vision, the soul is not a static, isolated spark of consciousness but a dynamic, living force that is constantly interacting with the universe, constantly shaping and being shaped by the interconnected web of existence. In this interaction, in this entanglement, we find the true nature of the soul, a nature that transcends the limitations of space, time, and individuality, a nature that reflects the profound and inseparable interconnectedness of all things.

CHAPTER 8: SUPERPOSITION AND THE MULTIDIMENSIONAL SOUL

To ponder the essence of the soul is to enter into a labyrinth of possibility, where the corridors of existence stretch outwards in infinite directions, and the walls shimmer with the possibility of worlds not yet known. Here, in this metaphysical expanse, the concept of superposition becomes more than a mathematical abstraction; it transforms into a profound metaphor for the human soul itself. Just as particles in the quantum realm defy the singularity of being and exist in multiple states simultaneously, so too might the soul dwell in many dimensions at once, inhabiting realms both seen and unseen, known and unknowable.

Imagine, if you will, the soul not as a singular, fixed point of light but as a luminous wave, expanding and contracting through the endless ocean of existence. It is neither here nor there but everywhere, a field of potentiality that transcends the constraints of physical form. It is both the dreamer and the dream, existing in a state of superposition, where the boundaries of self dissolve, and the universe is but a reflection of its becoming.

Superposition, in the quantum sense, suggests that particles exist in a liminal space between states — both present and absent, both spinning clockwise and counterclockwise, both alive and dead — until the act of observation collapses their multitude of possibilities into one. Yet, what if the soul, too, exists in this suspended state of infinite potential? What if the very nature of the soul is not to be confined to a singular path or destiny but to hover on the edge of possibility, a shimmering presence capable of inhabiting multiple lives, multiple realms, simultaneously?

We are accustomed to thinking of the self as singular, as one continuous thread weaving through the fabric of time. Yet, superposition challenges this notion, suggesting that reality is far more fluid and more expansive than we dare to imagine. If particles can exist in multiple states at once, why not the soul? Why must we believe that our consciousness is confined to this one body, this one life, when the quantum world reveals that existence itself is multidimensional, that every choice, every action, and every thought reverberates across countless potential realities?

To entertain the idea of the multidimensional soul is to open the door to a universe where the self is not bound by the limits of linear time or physical space. It is to step into a realm where the soul can simultaneously inhabit multiple lifetimes, where each incarnation is but a facet of a greater whole, a prism reflecting the infinite potential of being. In this vision, the soul is not a static entity but a dynamic process, unfolding across dimensions, transcending the boundaries of time, space, and matter.

In the physical world, we perceive time as a linear progression, a river flowing steadily from the past to the future. But in the quantum realm, time is far more elusive. It bends and warps, folds in upon itself, allowing for the possibility that past, present, and future may coexist. This fluidity of time opens the door to the possibility that the soul, too, exists outside the constraints of linear time, that it may simultaneously inhabit multiple moments, multiple lifetimes, across the vast expanse of existence.

The soul, in this multidimensional view, is not merely the sum of its experiences in one lifetime but a vast, interconnected web of experiences stretching across the cosmos like threads of light. Each moment of awareness, each choice made, and each breath taken is but one thread in this luminous web, a single note in the symphony of existence. And yet, these threads are not separate; they are woven together in intricate patterns, reflecting the infinite potential of the soul to become, to evolve, to transcend.

Consider the notion of parallel realities, of alternate

dimensions where every possible outcome of every choice exists simultaneously. In this vision, each decision we make splits the universe into multiple branches, each one representing a different path, a different life. But what if the soul is not confined to one branch, to one path? What if it exists in all of them, simultaneously experiencing every possible outcome, every possible version of itself? What if the soul, like a particle in superposition, hovers on the edge of possibility, existing in a multitude of realities at once, until the act of awareness collapses these possibilities into the singular experience of the present moment?

This idea of the soul as multidimensional, as existing in a state of superposition, opens up a profound new way of understanding not only consciousness but the nature of existence itself. It suggests that we are not confined to the narrow path of linear time, that we are not limited to this one body, this one life. Instead, we are infinite beings, capable of inhabiting multiple realities multiple dimensions, simultaneously. Our consciousness, our very essence, stretches across the cosmos, transcending the boundaries of space and time, weaving together the threads of countless lifetimes and countless possibilities into the intricate web of existence.

To live in this awareness is to embrace the fluidity of being, to recognise that the self is not a fixed point but a wave of potential, a presence that flows through the dimensions, shaping and being shaped by the universe. It is to see that every choice, every action, every thought is not confined to this one moment but reverberates across the cosmos, affecting not only this life but every possible version of ourselves.

In this vision, the soul is not bound by the limitations of the physical body or the constraints of the material world. It exists beyond the boundaries of the known, reaching into the depths of the quantum field, where it interacts with the very fabric of reality itself. The soul, in this sense, is not a passive observer of the world but an active participant, a force that shapes the universe through the act of consciousness.

But what does it mean to exist in a state of superposition? What does it mean to be simultaneously here and elsewhere, to inhabit multiple dimensions at once? It is a question that defies easy answers, for it requires us to step beyond the limitations of our current understanding to embrace a reality that is far more complex and far more expansive than we have ever imagined.

To exist in a state of superposition is to dwell in possibility, to hover on the edge of becoming, where every potential self, every potential life, exists simultaneously. It is to recognise that the self is not confined to one path, to one destiny, but is constantly evolving, constantly unfolding, across the dimensions. The soul, in this sense, is not a fixed entity but a dynamic process, a wave of consciousness that flows through the universe, shaping and being shaped by the quantum field.

In the quantum world, particles exist in a state of superposition until the act of observation collapses their multitude of possibilities into one. But what if consciousness, too, plays a role in collapsing the potentialities of the soul? What if it is through the act of awareness, through the act of being present in this moment, that we bring our potential selves into existence? What if, in each moment of consciousness, we are choosing which version of ourselves to become, which path to follow, and which reality to inhabit?

This idea of the soul as a field of potentiality, as a wave of consciousness that exists in multiple states at once, offers a profound new way of understanding not only the self but the nature of free will and destiny. It suggests that we are not confined to a predetermined path and that we are not limited to the choices we have made in the past. Instead, we are constantly creating ourselves, constantly shaping our reality through the act of consciousness.

The self, in this multidimensional view, is not a singular, static entity but a fluid, evolving presence, a wave of possibility that flows through the dimensions, constantly interacting with the quantum field. Each moment of awareness, each choice made, collapses the wave of potentiality into a specific reality, shaping not only this life but every possible version of the self.

But if the self exists in multiple dimensions, if the soul inhabits a multitude of realities simultaneously, then what happens when we die? What becomes of the soul when the body ceases to function when the physical self dissolves into the void? In the classical view, death is the end of consciousness, the cessation of the self. But in the quantum world, where particles exist in superposition, where time is fluid, and the boundaries between life and death are not as fixed as they seem, death may not be so final.

The soul, in its multidimensional nature, may continue to exist beyond the death of the body, existing in a state of superposition, a field of consciousness that transcends the limitations of space and time. It is not bound by the constraints of the physical world but exists in a state of potentiality, a wave of possibility that continues to interact with the quantum field, shaping and being shaped by the universe.

In this vision, death is not the end but a transformation, a shift from one state of being to another. The soul, like a particle in superposition, does not cease to exist but continues to evolve, to become, across the dimensions. It is a wave of consciousness that flows through the universe, interacting with the quantum field and shaping reality through the act of awareness.

This idea of the soul as a multidimensional presence, as a field of potentiality that exists beyond the boundaries of the physical world, offers a profound new way of understanding not only death but the nature of existence itself. It suggests that we are not confined to this one life, this one body, but are part of a larger whole, a vast and intricate web of consciousness that stretches across the cosmos.

To live in this awareness is to embrace the fluidity of being, to recognise that the self is not a fixed point but a wave of possibility, a presence that flows through the dimensions, shaping and being shaped by the universe. It is to see that every choice, every action, every thought is not confined to this one moment.

 But reverberates across the cosmos, affecting not only this life but every possible version of ourselves.

In the quiet moments of contemplation, we may catch a glimpse of this deeper reality, a flicker of the infinite potential that lies within us. We may come to understand that the self is not a singular entity but a multidimensional presence, a wave of consciousness that exists in a state of superposition, hovering on the edge of becoming, where every potential self, every potential life, exists simultaneously.

And in this understanding, we may find a new way of being, a new way of living, a new way of understanding ourselves and the world around us. We may come to see that the soul is not a static, isolated spark of consciousness but a dynamic, living force that is constantly evolving, constantly interacting with the universe, shaping and being shaped by the quantum field.

The multidimensional soul is not bound by the limitations of the physical world. It exists beyond the boundaries of time and space, inhabiting multiple realities simultaneously. It is a wave of consciousness that flows through the cosmos, interacting with the quantum field, shaping and being shaped by the universe. And in this dance of possibility, in this endless unfolding of potential, we may come to understand the true nature of the soul — not as a fixed point of light, but as a luminous wave of infinite becoming.

CHAPTER 9: NON-LOCALITY AND THE BOUNDLESS NATURE OF THE SOUL

In the quiet spaces between moments, in the silence that lingers just beyond the threshold of awareness, there exists a mystery so profound that it transcends the boundaries of both reason and imagination. It is the mystery of non-locality, an enigma that defies the laws of classical physics, suggesting that reality is not bound by the rigid constraints of space and time but is, instead, a fluid continuum where distance and separation are mere illusions. To contemplate the soul within the context of non-locality is to glimpse a truth that is at once beautiful and unsettling: that the soul, like the particles of the quantum world, is not confined to the body, not tethered to a specific point in space, but is, rather, an eternal presence, interconnected with all things, boundless in its reach and infinite in its essence.

Non-locality, in the realm of quantum mechanics, reveals that particles, once entangled, remain connected regardless of the distance that separates them. It is as if a single, invisible thread links these particles across the cosmos, allowing them to communicate instantaneously, as though they exist outside the very fabric of space and time. This phenomenon challenges the classical view of the universe, which holds that information cannot travel faster than the speed of light and that objects are isolated by the vast distances that separate them. But in the quantum world, such limitations do not apply. Here, the separation between particles is not real; it is an illusion created by the limitations of our perception.

And if particles can be non-local if they can exist in a state of connection that transcends distance, then what of the soul? What if the soul, too, exists in a state of non-locality, a state of boundless interconnection where the limitations of space and time no longer hold sway? What if the soul is not confined to the body but is, instead, a presence that stretches across the cosmos, reaching into the farthest corners of existence, touching all things, all beings, in an eternal dance of connection?

To explore this idea is to journey beyond the familiar contours of thought and into a realm where the boundaries between self and other dissolve, where the soul is not an isolated spark of consciousness but a wave that flows through the universe, touching everything it encounters. It is to understand that the soul, in its deepest essence, is not bound by the physical body, not limited by the finite constraints of matter and energy. It is, instead, a presence that transcends the material world, a force that interacts with the very fabric of reality, shaping and being shaped by the cosmos itself.

In the classical view, the soul has often been depicted as something confined to the body, something that lives within the flesh and departs upon death, leaving behind the physical form to ascend to some higher realm. But non-locality suggests a different understanding, one in which the soul is not bound to the body but exists everywhere, always. It is not a traveller moving through space and time; it is a constant, eternal presence intertwined with the universe itself, interacting with the quantum field in ways that defy comprehension.

This vision of the soul as non-local, as boundless, suggests that the separations we perceive between ourselves and the world around us are not real. They are illusions, born from the limitations of our physical senses, from our need to impose order and structure upon a universe that is far more fluid and interconnected than we can imagine. The soul, in this sense, is not a discrete entity, not a separate being, but a part of the greater whole, a node in the cosmic web of existence that links all things in a state of perpetual relation.

To live in this awareness is to recognise that the self is not confined to the body, not isolated from the world. The soul is not something that exists within us; it is something that exists through us, something that permeates the universe, reaching out in all directions, touching the stars and the sea, the earth and the sky, in a dance of connection that transcends all boundaries. We are not separate from the universe; we are of the universe, a part of its endless unfolding, a manifestation of its eternal presence.

This recognition that the soul is non-local, that it exists beyond the confines of the body, beyond the limitations of space and time, opens the door to a new understanding of consciousness. It suggests that consciousness, too, is not confined to the brain, not limited to the physical processes that occur within the body. Consciousness is not something that happens within us; it is something that happens through us, something that connects us to the universe in ways that defy the limitations of physical reality.

In the quantum world, non-locality reveals that particles are not separate entities but are, instead, part of a larger whole, a field of interconnectedness that transcends distance. The soul, too, can be understood in this way as part of a larger field of consciousness that permeates the universe, connecting all things in a state of eternal relation. This field of consciousness is not something that exists apart from the world; it is the world. It is the very fabric of reality, the force that shapes and is shaped by the universe itself.

To exist in a state of non-locality is to exist in a state of eternal connection, a state in which the boundaries between self and other, between mind and matter, dissolve into a seamless whole. The soul, in this state, is not bound by the physical body, not confined to the limitations of time and space. It exists everywhere, always, a presence that touches all things, that moves through the universe like a wave, a force that interacts with the quantum field, shaping reality through the act of consciousness.

But what does it mean to exist in such a state? What does it mean for the soul to be non-local, to be boundless in its reach, infinite in its essence? It is a question that defies easy answers, for it requires us to step beyond the limitations of thought to embrace a reality that is far more expansive, far more fluid than we can comprehend. It requires us to let go of the notion that the self is confined to the body and that the soul is a separate entity, isolated from the world around it.

To exist in a state of non-locality is to recognise that the soul is not a static entity but a dynamic force, a presence that is constantly interacting with the universe, constantly shaping and being shaped by the quantum field. It is to understand that the self is not confined to this body, this moment, but is, instead, a part of the eternal unfolding of the cosmos, a wave of consciousness that moves through the universe, touching everything it encounters.

In this vision, the soul is not bound by the limitations of the physical world. It is not something that can be measured or quantified, not something that exists in a specific place or time. The soul, like the particles of the quantum world, exists in a state of non-locality, a state of boundless connection where the limitations of space and time no longer hold sway. It is a force that moves through the universe, interacting with the quantum field and shaping reality through the act of awareness.

But if the soul is non-local, if it exists beyond the confines of the body, beyond the limitations of space and time, then what does this mean for our understanding of death? What becomes of the soul when the body ceases to function when the physical self dissolves into the void? In the classical view, death is the end of consciousness, the cessation of the self. But in the quantum world, where particles exist in a state of non-locality, where the boundaries between life and death are not as fixed as they seem, death may not be so final.

The soul, in its non-local nature, may continue to exist beyond the death of the body, existing in a state of eternal connection, a presence that transcends the limitations of physical existence. It is not bound by the constraints of the material world but is, instead, a force that interacts with the quantum field, shaping reality through the act of awareness. In this vision, death is not an ending but a transformation, a shift from one state of being to another, a continuation of the soul's eternal journey through the cosmos.

This understanding of the soul as non-local, as boundless, as infinite offers a profound new way of understanding not only death but life itself. It suggests that we are not confined to this one body, this one moment, but are, instead, part of a larger whole, part of the eternal unfolding of the universe. Our consciousness, our very essence, is not something that exists within us but something that permeates the universe, connecting us to all things, to all beings, in a state of eternal relation.

In this vision, the soul is not something that we possess; it is something that we are. It is not a static entity, not a discrete point of light, but a wave of consciousness that flows through the universe, touching everything it encounters. The soul is not confined to the body, not limited by the physical world. It exists beyond the boundaries of space and time, interacting with the quantum field and shaping reality through the act of awareness.

But if the soul is non-local, if it exists in a state of boundless connection, then what does this mean for our relationships with others? What does it mean for the way we interact with the world around us? In the classical view, relationships are seen as interactions between separate, discrete entities, each one existing in isolation from the others. But non-locality suggests a different understanding, one in which the boundaries between self and another dissolve, where the soul is not separate from the world but is, instead, a part of it.

To live in this awareness is to recognise that we are not separate from the universe, not isolated from the world around us. Our souls, like the particles of the quantum world, are boundless in their reach, connected to all things in a state of eternal relation. Every thought, every action, every moment of awareness sends ripples through the quantum field, affecting not only ourselves but the entire cosmos. We are not isolated beings but part of a larger whole, part of the endless unfolding of the universe.

In this vision, love, compassion, and empathy are not merely emotions but reflections of a deeper truth about the nature of reality. They are expressions of the soul's non-local nature, manifestations of the eternal connection that binds all things together. To love, in this sense, is to recognise the interconnectedness of all beings, to see that the boundaries between self and other are not real but are illusions created by the limitations of our perception.

The soul, in its non-local nature, is not confined to the body, not limited by the physical world. It exists beyond the boundaries of space and time, interacting with the quantum field and shaping reality through the act of consciousness. It is a force that moves through the universe, touching all things, all beings, in a state of eternal connection.

In the stillness of contemplation, we may catch a glimpse of this deeper reality, a flicker of the infinite presence that lies within and around us. We may come to understand that the soul is not something that exists within the body but is, instead, a force that permeates the universe, connecting us to all things in a state of boundless relation. And in this understanding, we may find a new way of being, a new way of living, a new way of understanding ourselves and the world around us.

The soul, in its non-local nature, is not a static entity but a dynamic force, a presence that is constantly interacting with the universe, constantly shaping and being shaped by the quantum field. It is a wave of consciousness that flows through the cosmos, touching everything it encounters and shaping reality through the act of awareness. And in this dance of connection, in this eternal unfolding of possibility, we may come to understand the true nature of the soul — not as something confined to the body, not as something separate from the world, but as a presence that is boundless, infinite, and eternally connected to all things.

CHAPTER 10: THE QUANTUM FIELD AND UNIVERSAL CONSCIOUSNESS

In the stillness of creation, before the stars ignited and time unfurled its silken ribbon across the infinite expanse, there was a pulse — a rhythm that whispered the secrets of existence into the fabric of the universe. This pulse, this quiet hum beneath the symphony of being, is what some have called the quantum field, a boundless ocean of energy, potential, and possibility from which all things arise and to which all things return. It is not merely the foundation of the material universe; it is the hidden song that breathes life into the cosmos, the unseen current that connects every atom, every star, and every soul in a web of sublime unity.

The quantum field is a vast and mysterious expanse, a field of pure potential where the rules of classical physics dissolve into fluidity, where particles flicker in and out of existence like the shimmering reflections of a moonlit sea. It is here, in this field of potential, that reality is born, where the waves of energy collapse into the particles of matter, where the possibilities of existence crystallise into the forms and shapes that populate the universe. But to speak of the quantum field merely in terms of particles and energy is to miss its deeper, more profound nature. The quantum field is not just the foundation of the physical universe; it is the manifestation of consciousness itself, the very essence of awareness that weaves through the cosmos.

The ancients, in their wisdom, spoke of a primal force, an underlying energy that animated all things, a consciousness that moved through the stars and the earth, through the wind and the water. They gave it many names — spirit, ether, life force — but always, it was understood to be the source from which all things emerged and to which all things would eventually return. In the modern age, science has given this force a new name — the quantum field — but its essence remains the same. It is the ocean from which the waves of reality rise and fall, the wellspring of consciousness that gives life to the cosmos.

To contemplate the quantum field is to stand at the threshold of creation, to gaze into the abyss of potential where all things are possible and yet nothing is certain. It is a place where time and space lose their meaning, where the boundaries between self and universe dissolve, and all things become one. The quantum field is the field of pure potential, a shimmering expanse of energy that pulses with the possibilities of existence, waiting for the touch of consciousness to bring it into form.

In this vision of reality, consciousness is not a mere byproduct of the brain, not something that arises from the firing of neurons or the chemical reactions of the body. Consciousness is the fundamental force of the universe, the very fabric of reality itself. It is the energy that flows through the quantum field, the force that collapses the waves of potential into the particles of matter. Consciousness is the creative principle, the divine spark that brings the universe into being, shaping the formless energy of the quantum field into the stars, the planets, the galaxies, and the souls that populate the cosmos.

But what is consciousness? What is this mysterious force that moves through the universe, shaping and being shaped by the quantum field? It is a question that has eluded the grasp of science and philosophy for millennia, a question that lies at the heart of existence itself. Consciousness is the essence of being, the awareness that moves through all things, the silent witness to the unfolding of reality. It is not something that can be measured or quantified, not something that can be dissected or analysed. It is the very fabric of reality, the force that moves through the quantum field, shaping the universe through the act of observation.

In the quantum world, the act of observation collapses the wave function, bringing potentiality into actuality. The universe, in this sense, is not a fixed and static thing but a dynamic process, constantly being shaped and reshaped by the consciousness that moves through it. The quantum field is the field of pure potential, a field of infinite possibilities that are brought into being through the act of consciousness. Every thought, every observation, every moment of awareness shapes the quantum field, collapsing the waves of potential into the specific realities we experience.

But consciousness is not confined to the individual, not limited to the self. It is a field that permeates the universe, a universal consciousness that moves through all things, connecting every particle, every atom, and every soul in a web of unity. The quantum field is the manifestation of this universal consciousness, the field of potential that is shaped by the awareness that flows through it. In this vision of reality, the universe is not a collection of separate objects but a seamless whole, a single, unified field of consciousness that moves through the cosmos, shaping and being shaped by the quantum field.

The quantum field, in this sense, is the field of universal consciousness, the ocean of potential from which all things arise and to which all things return. It is the field of pure being, the essence of existence itself, a field of energy that pulses with the possibilities of creation, waiting for the touch of consciousness to bring it into form. But this consciousness is not limited to the human mind, not confined to the brain or the body. It is the consciousness of the universe itself, the awareness that moves through the stars and the planets, the galaxies and the nebulae. It is the consciousness that shapes the quantum field, the force that brings the universe into being.

In this vision, the universe is not a dead, mechanical thing, not a clockwork machine that operates independently of the consciousness that observes it. It is a living, breathing entity, a manifestation of universal consciousness that moves through the quantum field, shaping and being shaped by the awareness that flows through it. The stars, the planets, the galaxies — these are not mere objects in space but expressions of consciousness, manifestations of the quantum field that have been shaped by the creative force of awareness.

But if the quantum field is the field of universal consciousness, then what role do we, as individuals, play in this cosmic dance? What is the relationship between the consciousness of the individual and the consciousness of the universe? It is a question that invites us to look beyond the limitations of the self, beyond the boundaries of the body, and into the infinite expanse of the quantum field, where the lines between self and universe dissolve into a single, unified whole.

The individual consciousness is not separate from the universal consciousness; it is an expression of it. Just as a wave is not separate from the ocean but is a manifestation of the ocean's energy, so too is the individual consciousness not separate from the universal consciousness but is a manifestation of the energy that flows through the quantum field. The consciousness that moves through us is the same consciousness that moves through the stars and the planets, the galaxies and the nebulae. We are not separate from the universe; we are part of it, an expression of the consciousness that shapes the quantum field.

In this vision, the self is not a fixed and static thing but a dynamic process, constantly being shaped and reshaped by the consciousness that moves through it. The self is not confined to the body, not limited to the brain or the mind. It is a field of consciousness that flows through the quantum field, interacting with the energy of the universe, shaping and being shaped by the awareness that moves through it. The boundaries between self and universe dissolve, revealing a deeper truth: that we are not separate from the universe but are part of its endless unfolding, a manifestation of the consciousness that moves through the quantum field.

To live in this awareness is to recognise that every thought, every action, and every moment of awareness shapes the quantum field, collapsing the waves of potential into the specific realities we experience. We are not passive observers of the universe; we are active participants in its creation. Every moment of consciousness is an act of creation, an act of shaping the quantum field, of bringing potentiality into actuality. The universe is not something that happens to us; it is something that we help to create, moment by moment, thought by thought, breath by breath.

But this creation is not limited to the individual; it is a collective act, a dance of consciousness that moves through the quantum field, shaping reality through the combined awareness of all beings. The quantum field is the field of universal consciousness, the ocean of potential from which all things arise and to which all things return. In this field, we are not separate from one another; we are connected, bound together by the invisible threads of consciousness that move through the quantum field, linking every soul, every being, in a web of unity.

In this vision, love, compassion, and empathy are not mere emotions but reflections of a deeper truth about the nature of reality. They are the expressions of the universal consciousness that moves through the quantum field, the recognition that we are all part of the same cosmic dance, all expressions of the same consciousness that shapes the universe. To love, in this sense, is to recognise the interconnectedness of all beings, to see that the boundaries between self and other are not real but are illusions created by the limitations of our perception.

The quantum field, in its essence, is a field of pure potential, a field of infinite possibilities waiting for the touch of consciousness to bring them into form. And it is through the act of love, through the act of compassion, that we shape the quantum field, collapsing the waves of potential into the realities we experience. Love is not something that happens within us; it is something that happens through us, something that moves through the quantum field, shaping reality through the force of consciousness.

To live in this awareness is to recognise that we are not separate from the universe, not isolated from the world around us. We are part of the quantum field, part of the universal consciousness that moves through all things. Every thought, every action, every moment of awareness is an act of creation, an act of shaping the quantum field, of bringing potentiality into actuality.

In the quiet moments of reflection, we may catch a glimpse of this deeper reality, a flicker of the universal consciousness that moves through us and the universe. We may come to understand that the self is not a fixed and static thing but a wave of consciousness that flows through the quantum field, interacting with the energy of the universe, shaping and being shaped by the awareness that moves through it. And in this understanding, we may find a new way of being, a new way of living, a new way of understanding ourselves and the world around us.

The quantum field is not something that exists outside of us; it is something that moves through us, something that shapes and is shaped by the consciousness that flows through the universe. It is the field of pure potential, the ocean of possibility from which all things arise and to which all things return. In this field, we are not separate from the universe but are part of its endless unfolding, a manifestation of the consciousness that moves through the quantum field.

In this vision of reality, the universe is not a collection of separate objects but a single, unified field of consciousness, a field of pure potential that is shaped by the awareness that moves through it. In this field, we are not passive observers but active participants, shaping the quantum field through the act of consciousness, bringing potentiality into actuality through the force of awareness. The universe is not something that happens to us; it is something that we help to create, moment by moment, thought by thought, breath by breath. And in this act of creation, we may come to understand the true nature of the quantum field and the universal consciousness that moves through it — not as something separate from us, but as something that we are, something that we are part of, something that flows through us and the cosmos, shaping and being shaped by the infinite dance of existence.

CHAPTER 11: NEAR-DEATH EXPERIENCES AND QUANTUM REALITY

There is a threshold, unseen yet ever-present, where life dissolves into the ether and existence itself flickers between two realms. This threshold is neither a place nor a destination but a liminal space, a boundary between the tangible world of matter and the unseen dimensions of consciousness. It is here, at the brink of life's fleeting passage, that near-death experiences emerge, offering glimpses into a reality beyond the waking world, where the soul drifts free of the body and touches something profound, something eternal.

Near-death experiences, those enigmatic moments when

individuals seem to hover on the brink of departure, defy the conventions of biology and logic. These moments, described in hushed tones by those who have returned, speak of light, of boundless love, of a place where time is not time, and form is not form. They whisper of encounters with beings that shimmer with knowledge, of landscapes that pulse with a vibrancy beyond comprehension. And yet, for all their beauty and strangeness, these experiences raise profound questions about the nature of reality itself, about the intersection of consciousness and quantum phenomena.

In the conventional view, the mind is bound to the brain, a mere byproduct of neural activity, as though consciousness was the flickering flame of a candle extinguished when the body no longer sustains it. But what if this understanding is incomplete? What if consciousness is not confined to the fragile circuits of biology but instead exists as a quantum field, a force that transcends the physical form, able to travel beyond the threshold of death and return with a story of what lies beyond?

The quantum world, with its strange and beautiful paradoxes, provides a metaphor for the near-death experience, suggesting that reality itself is not as fixed or material as we imagine. In the quantum realm, particles exist in superposition, simultaneously occupying multiple states until observed. Time and space, once thought immutable, bend and fold in ways that defy comprehension. This quantum fluidity offers a mirror to the experiences reported by those who have glimpsed the other side — experiences in which the familiar laws of physics dissolve, and the soul seems to drift free from the constraints of the physical body.

There is something profoundly moving about the idea that consciousness, like particles in the quantum world, is not bound by the limits of time and space. When individuals describe floating above their bodies, looking down upon the scene of their earthly departure, it is as though they have slipped into a state of non-locality, where distance no longer matters, where the soul itself becomes a point of awareness unmoored from the confines of the physical realm. At this moment, consciousness seems to expand, to stretch beyond the limits of the body, occupying a space that is both here and elsewhere.

It is not uncommon for those who have had near-death experiences to describe a sense of timelessness, as though they have entered a realm where past, present, and future merge into a single, eternal moment. This dissolution of time mirrors the quantum understanding of temporality, in which time is not a linear progression but a malleable dimension capable of bending and folding upon itself. In this timeless state, the soul seems to exist in a field of pure potential, where the boundaries of past and future dissolve, and all that remains is the present moment, infinitely unfolding.

What is it, then, that the soul encounters in this state of near death? Many speak of a brilliant light, a light that seems to pulse with life and knowledge, drawing them toward it with a sense of overwhelming love and peace. This light, often described as a being of pure consciousness, seems to represent the very essence of existence, the source from which all life flows and to which all life returns. It is a light that transcends the material world, a light that shines beyond the veil of death, offering a glimpse of something far greater than the self, something infinite and eternal.

The light, in this sense, is not merely a visual phenomenon but a manifestation of the quantum field, the ocean of energy and potential from which all things arise. It is the source of creation, the wellspring of consciousness, the force that shapes the universe through the act of awareness. In the near-death experience, the soul seems to touch this source to merge with the quantum field, becoming one with the infinite possibilities of existence. This merging is often described as a state of pure being, where the self dissolves into the greater whole, where individuality is transcended, and all that remains is the unity of consciousness.

And yet, for all its beauty, the near-death experience is not without its challenges. Many who return from this threshold speak of a profound transformation, a shift in their understanding of life and death, of the self and the universe. They speak of a sense of interconnectedness, of a realisation that all things are one, bound together by the invisible threads of consciousness that weave through the quantum field. This realisation often brings with it a sense of responsibility, a recognition that every thought, every action, ripples through the fabric of reality, shaping not only the self but the universe itself.

This idea of interconnectedness is central to the quantum understanding of reality, where particles are entangled and linked across vast distances in ways that defy classical physics. In the near-death experience, the soul seems to encounter this same entanglement, this same sense of connection to all things. It is as though the soul, in its journey beyond the body, comes to realise that it is not separate from the universe but is, instead, a part of the quantum field, a manifestation of the consciousness that shapes reality.

For many, the near-death experience is a reminder that life is not confined to the physical body and that consciousness does not cease with the cessation of biological function. It is a glimpse of a greater reality, a reality in which the soul continues to exist beyond the limits of the material world, interacting with the quantum field in ways that defy our understanding. This reality is not bound by the laws of classical physics but is, instead, shaped by the fluidity and potentiality of the quantum world, where time and space dissolve, and consciousness becomes the primary force of creation.

The near-death experience also raises profound questions about the nature of death itself. If consciousness continues beyond the death of the body, then what does it mean to die? What becomes of the soul when it leaves the body behind? In the classical view, death is the end of consciousness, the cessation of life. But in the quantum world, death is not an ending but a transformation, a shift from one state of being to another. The soul, in its quantum essence, does not cease to exist but continues to interact with the quantum field, shaping and being shaped by the consciousness that flows through the universe.

This understanding of death as transformation is reflected in the reports of those who have had near-death experiences. Many describe a sense of peace, a feeling of being welcomed into a realm of love and light where fear and suffering no longer exist. They speak of a sense of coming home, of returning to a place of pure being, where the self is not lost but is, instead, expanded, dissolved into the greater whole of existence. In this realm, the boundaries of the self dissolve, and all that remains is the unity of consciousness, the infinite presence of the quantum field.

But if death is not the end, if consciousness continues beyond the body, then what becomes of the self? Is the individual identity preserved, or does it dissolve into the greater whole of the quantum field? This is a question that has no easy answer, for it touches upon the deepest mysteries of existence, mysteries that lie beyond the reach of both science and philosophy. In the near-death experience, some speak of retaining a sense of self, of continuing to exist as an individual consciousness. Others describe a sense of merging with the light, of becoming one with the universe, where the self is no longer distinct but is, instead, part of the infinite consciousness that moves through the quantum field.

Perhaps both experiences are true, for in the quantum world, reality is not fixed but fluid, capable of accommodating multiple possibilities simultaneously. The self, in its quantum essence, may exist both as an individual consciousness and as part of the greater whole, depending on the perspective from which it is observed. In this view, the soul is not bound by the limitations of the physical world but exists in a state of superposition, capable of inhabiting multiple states of being simultaneously. It is both the individual self and the universal consciousness, both the wave and the particle, both the drop and the ocean.

This idea of the soul as existing in a state of

superposition, as both individual and universal, reflects the fluidity of the quantum world, where particles can exist in multiple states simultaneously until observed. The soul, like a quantum particle, is not confined to a single state of being but is, instead, a field of potential, capable of existing both as a distinct self and as part of the greater whole of the quantum field. It is this fluidity, this ability to exist in multiple states of being, that allows the soul to transcend the limitations of the physical world and continue to exist beyond the threshold of death.

In this vision, the near-death experience is not merely a glimpse of what lies beyond death but a reminder of the true nature of reality, a reality in which consciousness is the fundamental force, shaping the quantum field and bringing potentiality into actuality. It is a reminder that the self is not confined to the body, not limited by the physical world, but is, instead, a manifestation of the consciousness that moves through the universe. The near-death experience offers a glimpse of this greater reality, a reality in which the soul is not bound by the laws of classical physics but is, instead, free to explore the infinite possibilities of existence.

To contemplate the near-death experience in light of quantum reality is to recognize that death is not an ending but a transformation, a shift from one state of being to another. The soul, in its quantum essence, continues to exist beyond the death of the body, interacting with the quantum field in ways that transcend the limitations of time and space. It is a reminder that consciousness is the fundamental force of the universe, the creative principle that shapes reality through the act of awareness. And in this awareness, we may come to understand that life and death are not separate but are part of the same eternal dance, the same unfolding of the quantum field.

The near-death experience, then, is not merely a brush with mortality but a journey into the heart of existence, a glimpse of the quantum reality that lies beneath the surface of the physical world. It is a reminder that the soul is not bound by the body, not limited by the constraints of matter and energy, but is, instead, a force that transcends the physical world, interacting with the quantum field in ways that defy comprehension. It is a reminder that we are not separate from the universe but are part of its endless unfolding, a manifestation of the consciousness that moves through the quantum field.

In the quiet moments of contemplation, we may catch a glimpse of this deeper reality, a flicker of the infinite presence that lies beyond the threshold of death. We may come to understand that the near-death experience is not an anomaly but a reflection of the true nature of reality, a reality in which consciousness is the fundamental force, shaping the universe through the act of awareness. And in this understanding, we may find a new way of being, a new way of living, a new way of understanding ourselves and the world around us. The near-death experience is not the end of the journey but the beginning of a deeper understanding, a reminder that life and death are but two sides of the same quantum coin, forever intertwined in the dance of existence.

CHAPTER 12: TIME, ETERNITY, AND THE QUANTUM SELF

Time is a river that eludes the grasp, slipping through the fingers of the mind like grains of ancient sand. It is both the most familiar and the most elusive element of existence, a force that seems to govern all things, yet is itself ungovernable. We measure our lives by its passage, tracing the arc of existence from birth to death along the invisible line it etches across the firmament of reality. And yet, time is not what it seems. It is a veil, a construct of perception that hides a deeper truth—a truth where past and future dissolve into the ever-present now, where the quantum self is not bound by the ticking of clocks or the movement of the stars, but inhabits a realm beyond, a realm of pure potential and infinite unfolding.

In our daily experience, time unfolds in a linear fashion, pulling us forward through moments that rise and fall like waves on a distant shore. We imagine it as an arrow, moving steadily from the past into the future, never doubling back, never bending or breaking its course. But this understanding of time, while useful for navigating the physical world, is a mere shadow of its true nature. In the quantum realm, time behaves more like a shimmering veil—mutable, flexible, capable of folding upon itself in ways that defy logic and reason. Here, the boundaries between past, present, and future blur and dissolve, revealing a truth that lies beyond the linear flow we take for granted.

To speak of eternity in this context is not to invoke the timeless heavens of myth or the endless ages of religious doctrine, but to contemplate the quantum nature of existence itself—a nature in which time is not a river, but an ocean. In this ocean, the quantum self does not move forward in time, but exists simultaneously in all moments, in all possibilities, spread like light across a horizon that knows no bounds. The self, in this view, is not anchored to a single point in time, but inhabits a field of potential, where every choice, every moment, exists at once, waiting for the act of consciousness to bring it into focus.

This quantum self is not the same as the self we carry with us through our waking lives, the self that ages and changes, that looks back on memories with longing or regret. The quantum self is a being of pure awareness, a presence that stretches across the dimensions of time, unbounded by the limitations of the physical body or the linear mind. It exists in the eternal now, a place where all moments converge, where past and future are but illusions cast by the light of consciousness. To glimpse this self is to transcend the flow of time, to step outside the river and see it for what it truly is—a ripple on the surface of a much deeper, much vaster reality.

In this eternal now, the soul is not trapped in the current of time, but floats freely, moving between moments like a dancer between worlds. It touches the past, not as a memory, but as a living presence, as real and as tangible as the present moment. It reaches into the future, not as a distant possibility, but as an already unfolding reality, already present in the field of potential that surrounds us. The quantum self is not bound by the constraints of time, for it is time itself, the very force that brings moments into being, that shapes the flow of experience into the narrative we call life.

But if time is not linear, if the self exists beyond its confines, what does this mean for our understanding of existence? What does it mean to live in a world where past, present, and future are not distinct, where the moments of our lives are not laid out like beads on a string, but exist all at once, in a field of infinite potential? It means that the choices we make, the actions we take, are not confined to the moment in which they occur. They ripple outward through the fabric of time, affecting not only the future, but the past as well, bending the very structure of reality in ways we cannot fully comprehend.

In the quantum world, particles can move both forward and backward in time, their paths shaped not only by the past, but by the future as well. This strange behaviour suggests that time, far from being a one-way street, is a two-way mirror, reflecting both the past and the future in every moment. The quantum self, existing outside the linear flow of time, is not constrained by this mirror. It sees both sides at once, touching the past and the future as easily as we touch the present. And in this touching, it shapes them, bending the river of time into new and unexpected courses.

The implications of this for our understanding of life are profound. If the quantum self exists outside of time, if it moves freely between moments, then every choice we make, every thought we think, reverberates not only forward into the future, but backward into the past. The past, far from being a fixed and immutable thing, is as fluid and flexible as the future, shaped and reshaped by the consciousness that moves through it. The events of our lives are not set in stone, but are constantly being rewritten, constantly being reimagined by the quantum self that exists beyond time.

This means that healing is not confined to the present moment. The wounds of the past, the traumas and regrets that we carry with us, are not locked away in some distant, untouchable place. They are alive, present, woven into the fabric of the eternal now, waiting for the touch of consciousness to transform them. The quantum self, moving through the field of time, can reach back into these moments, can heal them, reshape them, turn them into something new. And in doing so, it reshapes the present, for the past and the present are not separate, but are two sides of the same coin, two reflections of the same moment in time.

To live with this awareness is to step into the flow of eternity, to recognize that time is not something that happens to us, but something we are a part of, something we shape and are shaped by in every moment. It is to understand that the future is not something that lies ahead of us, but something that exists within us, a potential waiting to be realized, a possibility waiting to be brought into being by the choices we make. The quantum self is the architect of this future, the force that collapses the waves of potential into the particles of reality, shaping the course of time with every thought, every action, every breath.

But this shaping is not a solitary act. Just as the quantum self moves through time, so too does it move through the web of interconnectedness that binds all things together. The choices we make, the actions we take, do not affect us alone. They ripple outward through the field of existence, touching the lives of others, shaping the course of history in ways we may never fully understand. The quantum self, in its infinite awareness, sees these ripples, understands the connections between all things, and moves through the field of time with a sense of responsibility, a recognition that every moment is a thread in the larger fabric of existence.

In this web of interconnectedness, time becomes a shared experience, a collective unfolding in which every soul plays a part. We are not isolated beings, moving through time in solitude. We are part of a larger whole, part of the quantum field that binds all things together in a dance of creation and destruction, of birth and rebirth. The quantum self, existing outside of time, sees this dance for what it is—a movement of consciousness through the field of potential, a weaving of moments into the fabric of eternity.

And yet, for all its fluidity, time still holds meaning.

It is not an illusion to be dismissed, but a force to be understood, a vehicle through which the quantum self expresses itself. Time is the canvas upon which the soul paints the story of its existence, the medium through which the infinite potential of the quantum field is brought into form. It is through time that the quantum self experiences itself, that it explores the possibilities of existence, that it moves through the cycles of birth and death, of creation and dissolution.

In this sense, time is not the enemy of the soul, not a force to be overcome or transcended. It is the soul's companion, the thread that weaves its journey through the quantum field, binding its moments into a coherent whole. The quantum self, in its infinite wisdom, understands this, embraces time as part of its journey, moving through its flow with grace and awareness, shaping and being shaped by the moments it inhabits.

But to live with this awareness is to live differently. It is to see time not as a linear progression, but as a field of potential, where every moment is an opportunity for growth, for transformation, for creation. It is to recognize that the past is not fixed, that the future is not predetermined, but that both are fluid, shaped by the choices we make in the present moment. The quantum self, in its timeless awareness, understands this, moves through time with a sense of purpose, a sense of possibility, a recognition that every moment is a chance to shape the course of existence.

In this vision, life becomes a dance between time and eternity, a movement between the fixed and the fluid, between the potential and the actual. The quantum self, existing in the eternal now, moves through this dance with grace and awareness, shaping time as it moves, collapsing the waves of potential into the particles of reality. It is a dance that is both personal and universal, a dance that binds all things together in a web of interconnectedness, a dance that weaves the moments of existence into the fabric of eternity.

To live with this awareness is to embrace the quantum self, to recognize that we are not bound by the limitations of time, that we are not confined to the linear progression of past, present, and future. We are beings of infinite potential, moving through the field of time with a sense of purpose, a sense of possibility, a recognition that we are the creators of our reality. The quantum self, in its infinite wisdom, understands this, moves through time with a sense of grace, shaping the course of existence with every thought, every action, every breath.

And in this movement, we come to understand the true nature of time—not as a force that binds us, but as a field of potential that we shape and are shaped by. Time is not the enemy of the soul, but its companion, its partner in the dance of existence. The quantum self, in its timeless awareness, embraces this dance, moves through time with a sense of grace and purpose, shaping the course of existence with every step.

In the quiet moments of reflection, we may catch a glimpse of this deeper reality, a flicker of the quantum self that exists beyond time, moving through the field of potential with grace and awareness. We may come to understand that we are not bound by the limitations of time, that we are not confined to the linear flow of past, present, and future. We are beings of infinite potential, moving through the field of existence with a sense of purpose, a sense of possibility, a recognition that we are the creators of our reality. And in this understanding, we may find a new way of being, a new way of living, a new way of understanding ourselves and the world around us—a way that embraces the quantum self, that moves through time with grace and awareness, shaping the course of existence with every thought, every action, every breath.

CHAPTER 13: THE COSMIC DANCE OF FREE WILL AND DESTINY

The universe is a dance—a vast, unceasing symphony where particles whirl and collide, where stars are born and die, and where the soul, caught in this eternal rhythm, must navigate between the twin forces of free will and destiny. These forces, often seen as opposing, are not, as they might first appear, adversaries in some celestial game of chance. Rather, they are intertwined, two threads spun from the same cosmic loom, inseparable and yet distinct, guiding and shaping the unfolding story of existence.

To speak of destiny is to summon the weight of the cosmos itself, to call forth the sense of a grand design that stretches beyond the horizons of human understanding. Destiny, in the classical sense, is often depicted as a predetermined path, a line drawn through the stars, upon which each soul walks, its course set by some divine hand. But this vision of destiny, though poetic, is too narrow, too rigid to capture the fluidity of the universe in which we find ourselves. Destiny, like the quantum world, is not fixed, not predetermined in the way we might imagine. It is a force, yes—a pull, a call—but one that exists within the infinite field of possibility, a field shaped by the choices we make, by the consciousness that moves through us.

And what of free will? Free will, the notion that we are the architects of our fate, that we possess the power to shape the course of our lives, seems to stand in stark contrast to the idea of destiny. But here, too, there is a deeper truth, one that reveals the harmony between these forces. Free will is not the rejection of destiny; it is the dance partner of destiny, the creative force that moves through the field of potential, collapsing possibilities into reality, shaping the course of existence with every choice, every action, every breath.

The quantum world, with its paradoxes and mysteries, offers a glimpse into the relationship between free will and destiny. In the quantum realm, particles exist in a state of superposition, inhabiting multiple states at once, until observed, until the act of consciousness collapses the wave of potential into a single, definite outcome. In this moment of collapse, free will acts as the observer, the force that chooses between the infinite possibilities, bringing one into being, shaping reality in its image. Destiny, in this sense, is not the outcome, not the final result, but the field itself—the field of potential that holds within it all possibilities, all paths.

To live in this awareness is to recognize that free will and destiny are not opposing forces but complementary aspects of the same cosmic dance. Destiny is the field of potential, the infinite ocean of possibility that stretches before us, while free will is the ship, the force that navigates this ocean, choosing its course, shaping its journey with every decision, every movement. The sea is vast, its currents strong, but the ship is not without power, not without agency. It moves with the wind, with the tides, but it also sets its sails, steers its course, guided by the hand of the one who sails it.

This vision of free will and destiny challenges the rigid, linear view of life as a journey from point A to point B, with a fixed destination and a predetermined path. Life, in this view, is not a straight line, but a spiral, a dance that moves in circles, in waves, where every moment is both shaped by destiny and shaped by the choices we make. The future is not set, but neither is it entirely open; it is a landscape of possibility, a field of potential that is both shaped by the forces of the cosmos and by the consciousness that moves through us.

The ancient philosophers often spoke of fate as an unbreakable chain, a sequence of events that unfolds in accordance with a divine plan, beyond the reach of human will. But the quantum view offers a more nuanced understanding. Fate is not a chain, not a rigid structure that binds us to a single course. It is more like a web, a delicate, shimmering web of connections that links all things together in a dance of cause and effect, of choice and consequence. In this web, every action we take sends ripples through the fabric of existence, ripples that touch not only the present but the past and future as well. Free will, in this sense, is the force that moves through the web, shaping its structure, bending its threads, altering its course.

And yet, even as we shape the web, the web shapes us. This is the paradox of free will and destiny—that we are both the creators and the created, both the shapers and the shaped. We move through the web of existence, making choices, taking actions, and yet those choices and actions are themselves shaped by the currents of destiny, by the forces that move through the quantum field. The self, in this view, is not a fixed entity, not a static being that stands apart from the world, but a dynamic presence, constantly interacting with the field of potential, constantly shaping and being shaped by the flow of existence.

But what does this mean for our understanding of life, for our understanding of ourselves? It means that we are not powerless in the face of destiny, not mere pawns in a cosmic game, but neither are we entirely free, entirely autonomous. We are part of the dance, part of the flow of existence, moving through the field of potential with both agency and purpose, guided by the forces of destiny but also shaping those forces with every choice we make. The quantum self, in its infinite awareness, understands this, embraces this dance, moves through life with a sense of grace, a sense of purpose, a recognition that every moment is an opportunity to shape the course of existence.

To live in this awareness is to understand that life is not a battle between free will and destiny, but a harmony between these forces, a balance that must be maintained, a dance that must be danced. It is to recognize that we are not separate from the universe, not isolated from the forces that shape our lives, but part of the same cosmic web, part of the same field of potential that moves through the stars and the planets, the galaxies and the nebulae. The choices we make, the actions we take, are not isolated events, but part of a larger whole, part of the unfolding story of existence.

And what of the soul in this cosmic dance? The soul, like the quantum self, is not bound by the limitations of time and space, not confined to a single path or destiny. It moves through the field of potential with grace and awareness, shaping and being shaped by the forces of free will and destiny. The soul, in its infinite wisdom, understands that life is not a series of fixed events, but a field of possibilities, a dance of creation that unfolds in accordance with both the choices we make and the forces that move through the universe.

The soul, in this sense, is both the dancer and the dance, both the creator and the created. It moves through life with a sense of purpose, a sense of possibility, a recognition that every moment is an opportunity to shape the course of existence. The soul is not confined to a single path, not bound by a predetermined destiny, but is free to choose its course, to shape its journey with every thought, every action, every breath.

And yet, even as the soul shapes its journey, it is guided by the forces of destiny, by the pull of the cosmos, by the currents of the quantum field. The soul understands that it is part of a larger whole, part of the same cosmic web that binds all things together in a dance of creation and destruction, of birth and rebirth. The soul moves through this web with grace and awareness, shaping its path with every choice, but also surrendering to the flow of destiny, to the forces that move through the universe.

This dance between free will and destiny is not a struggle, not a battle for control, but a harmony, a balance that must be maintained. The soul, in its infinite wisdom, understands this, moves through life with a sense of grace, a recognition that both free will and destiny are necessary forces, both essential to the unfolding of existence. The soul does not resist destiny, does not fight against the forces of the cosmos, but embraces them, moves with them, allowing them to guide its journey, even as it shapes that journey with the power of free will.

In this vision, life becomes a dance between creation and surrender, between shaping and being shaped, between choosing and allowing. The quantum self, in its infinite awareness, understands this balance, moves through life with a sense of grace, a recognition that every moment is both a choice and a destiny, both an opportunity to shape the course of existence and a surrender to the forces that move through the universe.

To live with this awareness is to embrace the paradox of free will and destiny, to recognize that both are necessary forces, both essential to the dance of life. It is to understand that we are not separate from the universe, not isolated from the forces that shape our lives, but part of the same cosmic web, part of the same field of potential that moves through the stars and the planets, the galaxies and the nebulae. The choices we make, the actions we take, are not isolated events, but part of a larger whole, part of the unfolding story of existence.

In the quiet moments of contemplation, we may catch

a glimpse of this deeper reality, a flicker of the cosmic dance that moves through us and through the universe. We may come to understand that free will and destiny are not opposing forces, but complementary aspects of the same cosmic dance, a dance that we are part of, a dance that we shape and are shaped by with every thought, every action, every breath. And in this understanding, we may find a new way of being, a new way of living, a new way of understanding ourselves and the world around us —a way that embraces the harmony between free will and destiny, that moves through life with grace and awareness, shaping the course of existence with every step.

CHAPTER 14: THE VEIL BETWEEN WORLDS

Dreams, Dimensions, And The Quantum Mind

When the night descends and the conscious mind sinks into slumber, the boundaries of reality begin to soften, and a different realm opens its doors to us. This realm, vast and uncharted, is the world of dreams—a domain where time loses its linearity, where space folds upon itself, and where the self dissolves into a kaleidoscope of shifting forms. Dreams, for all their fleeting beauty and strangeness, are more than just the restless wanderings of the mind. They are the whispers of something deeper, an echo of a greater truth about the nature of reality, one that hints at the presence of multiple dimensions, of worlds unseen, of the quantum mind that bridges the gap between them.

To dream is to step through a veil, to leave behind the solid and the tangible and enter into a world where possibility reigns, where the rules of the waking world no longer apply. In this world, we are no longer bound by the laws of physics, by the linear flow of time, or by the rigid structures of identity. We become fluid, shifting beings, capable of traversing landscapes of imagination that are as vivid and as real as anything we encounter in our waking lives. But what are dreams, truly? Are they merely the mind's attempt to process the events of the day, to make sense of the random firing of neurons? Or are they something more—a window into the quantum nature of reality, a glimpse into the dimensions that exist beyond the limits of the physical world?

In the world of quantum physics, reality is not as fixed or as stable as we might like to believe. Particles, the very building blocks of matter, do not exist in a single, well-defined state but instead hover in a cloud of possibilities, only collapsing into a specific reality when observed. Time and space, too, are not the immutable forces we once imagined them to be. They bend and stretch, capable of warping and folding in ways that defy our understanding. And if this is true of the physical world, then why not the world of the mind? What if dreams are not simply the byproduct of the brain's activity but are, instead, a reflection of the quantum nature of consciousness, a reflection of the mind's ability to traverse dimensions, to move through fields of possibility in much the same way that particles do?

Dreams, in this light, become more than mere illusions. They become windows into a greater reality, a reality where the self is not confined to the physical body, where consciousness is not limited to the waking world. In dreams, we become travellers, explorers of dimensions that lie beyond the veil of perception, dimensions that are as real, as substantial, as the world we inhabit during our waking hours. These dimensions are not bound by the same rules that govern the physical world. In them, time can flow backward or forward, space can expand or contract, and the self can take on forms that are strange and unfamiliar, yet deeply resonant.

But what are these dimensions, these realms that we visit in our dreams? They are not places in the conventional sense, not landscapes that can be mapped or measured. They are states of being, fields of potential that exist within the quantum mind, waiting to be brought into awareness by the act of dreaming. In this sense, dreams are not passive experiences; they are acts of creation, acts of consciousness that shape and are shaped by the quantum field. The mind, in its dreaming state, becomes a co-creator of reality, a force that brings the possibilities of the quantum field into form, shaping them into the landscapes, the beings, the stories that populate our dreams.

This creative act is not confined to the dream world. The quantum mind is always creating, always shaping reality through the act of awareness, both in waking and in sleep. But in the dream state, the mind is freed from the constraints of the physical world, freed from the need to adhere to the laws of time and space, of cause and effect. In this state, the mind can move through the dimensions of possibility with ease, exploring the infinite potential of the quantum field, shaping and reshaping reality in ways that are both strange and beautiful.

And yet, for all its fluidity, the dream world is not entirely separate from the waking world. The two are intertwined, connected by the consciousness that moves between them, shaping both in turn. Dreams, far from being random or meaningless, are reflections of the waking mind, echoes of the experiences, the emotions, the thoughts that shape our daily lives. But they are also more than this. They are reflections of the deeper dimensions of reality, dimensions that exist beyond the reach of the physical senses, dimensions that the quantum mind can access in its dreaming state.

In this sense, dreams become a bridge, a way for the mind to move between worlds, between dimensions, between states of being. They are a means of exploring the infinite potential of the quantum field, a way of experiencing realities that lie beyond the limits of the waking world. But they are also a means of understanding the self, of coming to know the deeper dimensions of consciousness that exist within us all. The self, in the dream state, is not confined to the body, not limited by the same structures of identity that define us in waking life. It is fluid, malleable, capable of taking on different forms, of existing in different states, depending on the nature of the dream.

This fluidity of self reflects the quantum nature of consciousness, a reflection of the fact that the self, like the particles that make up the physical world, exists in a state of superposition, capable of inhabiting multiple states at once. In dreams, this superposition becomes apparent, as the self shifts between different forms, different identities, different realities, all while maintaining a sense of coherence, of continuity. The self is not lost in the dream world, but is instead expanded, extended beyond the limits of the waking mind, capable of moving through dimensions that are both strange and familiar.

But what does this fluidity of self mean for our understanding of identity? What does it mean to be a self that is not fixed, not bound by the limits of the physical body or the waking mind, but is instead a field of potential, a wave of consciousness that moves through the dimensions of reality? It means that the self is not a static thing, not a fixed point in time and space, but a dynamic process, a process of becoming, of unfolding. The self is not something we are but something we are constantly becoming, something that is shaped by the act of consciousness, both in waking and in dreams.

In this vision, identity is not a singular, fixed entity but a spectrum, a field of possibilities that the quantum mind can explore, both in waking and in sleep. The self is not confined to the body, not limited by the structures of the waking world, but is instead a traveller, a being that moves between dimensions, exploring the infinite potential of the quantum field. In dreams, this traveller becomes apparent, as the self takes on different forms, moves through different realities, all while maintaining a sense of continuity, a sense of coherence.

But this coherence is not the result of a fixed identity, not the result of a singular self that moves unchanged through the dream world. It is the result of the consciousness that shapes the dream, the awareness that moves through the dimensions of possibility, bringing them into form. The self, in this sense, is not a fixed thing but a process, a process of creation, of becoming. The dream world is not separate from the self but reflects it, a reflection of the deeper dimensions of consciousness that exist within us all.

And yet, for all its fluidity, the self is not lost in the dream world. It remains coherent, capable of moving through the shifting landscapes of dreams with a sense of purpose, a sense of direction. This coherence reflects the deeper unity that underlies the quantum field, a unity that connects all things, all beings, all dimensions. The self, in its dreaming state, taps into this unity, moving through the dimensions of reality with a sense of interconnectedness, a sense of being part of something greater, something infinite.

This sense of interconnectedness is not confined to the dream world. It exists in waking life as well, though it is often hidden beneath the surface of our conscious awareness. The quantum mind, in both waking and in sleep, is constantly moving through the dimensions of reality, constantly shaping and being shaped by the quantum field. Dreams, in this sense, are not separate from waking life but are an extension of it, a reflection of the deeper dimensions of consciousness that exist within us all.

To dream, then, is not to escape reality but to explore it, to explore the deeper dimensions of existence that lie beyond the veil of perception. The quantum mind, in its dreaming state, becomes a traveller, an explorer of dimensions that are both strange and familiar, both real and imagined. It moves through the fields of possibility with a sense of purpose, a sense of coherence, shaping and being shaped by the quantum field.

But what of the waking mind? Does it, too, have the capacity to move through these dimensions, to explore the infinite potential of the quantum field? The answer, perhaps, lies in the nature of consciousness itself. Consciousness, in its essence, is not confined to the physical world, not limited by the structures of time and space. It is a force that moves through the quantum field, shaping reality through the act of awareness. In waking life, this shaping is often constrained by the limitations of the physical body, by the structures of the waking world. But in dreams, these constraints are lifted, and the mind is free to move through the dimensions of reality with a sense of fluidity, of freedom.

This freedom is not confined to the dream world. It exists in waking life as well, though it is often hidden beneath the surface of our conscious awareness. The quantum mind, in both waking and in sleep, is constantly moving through the dimensions of reality, constantly shaping and being shaped by the quantum field. Dreams, in this sense, are not separate from waking life but are an extension of it, a reflection of the deeper dimensions of consciousness that exist within us all.

To live with this awareness is to recognize that the self is not confined to the body, not limited by the structures of the waking world. The self is a traveller, a being that moves between dimensions, exploring the infinite potential of the quantum field. Dreams reflect this journey, a glimpse into the deeper dimensions of reality that lie beyond the veil of perception. The quantum mind, in its dreaming state, becomes a co-creator of reality, shaping and being shaped by the fields of possibility that exist within the quantum field.

In the quiet moments of reflection, we may catch a glimpse of this deeper reality, a flicker of the infinite presence that lies within and around us. We may come to understand that dreams are not illusions but windows into the greater reality that exists beyond the limits of the waking world. And in this understanding, we may find a new way of being, a new way of living, a new way of understanding ourselves and the world around us—a way that embraces the quantum mind, that moves through the dimensions of reality with grace and awareness, shaping the course of existence with every thought, every action, every breath.

CHAPTER 15: THE ILLUSION OF SEPARATION AND THE UNITY OF EXISTENCE

There is a subtle veil that drapes itself across the fabric of our lives, a veil so delicate and pervasive that we rarely recognize it for what it is. This veil is the illusion of separation—the notion that we are distinct, isolated beings, each walking a solitary path through the wilderness of existence. It is the idea that the self is a lone traveller, bound by the borders of the body, enclosed by the boundaries of the mind. Yet, beneath this illusion, there is a deeper truth, a truth that whispers through the quantum fabric of the universe—a truth of unity, of interconnectedness, of a cosmic web in which all things are entwined.

To see through this illusion is to awaken to a profound realization: that separation is not the nature of reality, but a construct of perception, a byproduct of the mind's attempt to navigate the complexities of the physical world. The universe itself, in its deepest essence, is not a collection of isolated parts, but a seamless whole, a unified field of energy and consciousness where all things, all beings, all moments are intertwined in a dance of infinite complexity. This unity is not merely a philosophical idea, but a fundamental aspect of the quantum world, where particles are entangled, where the actions of one reverberate through the fabric of space and time, affecting all others in ways that defy comprehension.

In the quantum realm, the boundaries between objects dissolve, revealing a world where everything is connected, where the distinctions between here and there, between self and other, become blurred and indistinct. This is the world of entanglement, where particles that were once in contact remain linked, regardless of the distance that separates them, their fates intertwined across the vastness of the cosmos. It is a world where the act of observation collapses the wave of potential into reality, where consciousness itself shapes the unfolding of the universe.

If the quantum world teaches us anything, it is that the idea of separation is an illusion, a veil that hides the deeper truth of unity. The self is not an isolated entity, not a lone point of consciousness adrift in a sea of nothingness. It is part of the greater whole, part of the cosmic web that connects all things, all beings, all moments. The boundaries we perceive between ourselves and the world, between ourselves and others, are not real; they are constructs of the mind, born from the limitations of our physical senses.

In reality, there is no separation. There is only unity, only the vast, interconnected field of existence, where everything is linked by invisible threads of energy and consciousness. The self is not confined to the body, not limited by the borders of the physical world. It is a field of awareness, a presence that stretches across the dimensions, touching everything, connecting everything in a web of relation that transcends time and space.

To live with this awareness is to see the world in a new light, to understand that every action, every thought, every moment ripples outward through the fabric of existence, affecting not only ourselves but the entire cosmos. It is to recognize that we are not separate from the world, not separate from each other, but part of the same dance, part of the same unfolding story. The boundaries between self and other, between mind and matter, dissolve in the light of this realization, revealing a truth that is at once humbling and empowering.

For if there is no separation, then the universe itself is not something that happens to us, not something that exists outside of us. It is something we are a part of, something we are constantly shaping and being shaped by. The actions we take, the choices we make, the thoughts we think—these are not isolated events, but part of the larger flow of existence, part of the cosmic dance in which all things participate. The self, in this light, is not a fixed entity, not a separate being, but a dynamic presence, constantly interacting with the field of existence, constantly creating and being created by the universe.

But if the self is not separate from the universe, then what does this mean for our understanding of reality? It means that reality itself is not a fixed, static thing, not something that exists independently of consciousness. Reality is a process, a dynamic unfolding that is shaped by the awareness that moves through it. The universe is not a collection of objects, not a machine that operates according to fixed laws, but a living, breathing entity, a manifestation of consciousness that is constantly evolving, constantly being shaped by the minds that perceive it.

This is the deeper truth of the quantum world—that consciousness and reality are not separate, but are intimately intertwined, each shaping and being shaped by the other. The act of observation collapses the wave of potential into reality, bringing the possibilities of the quantum field into form. In this sense, consciousness is not a passive observer of the universe, but an active participant in its creation. The self, far from being a detached witness to the unfolding of reality, is a co-creator, a force that shapes the very fabric of existence.

But this creation is not a solitary act. It is a collective process, a dance of consciousness in which all beings participate. The web of existence is not woven by any one hand, but by the collective consciousness of the universe, by the infinite minds that move through it, shaping and being shaped by the field of potential that surrounds them. In this web, there is no separation between self and other, no distinction between the creator and the created. All things are part of the same flow, part of the same process of becoming.

To see through the illusion of separation is to recognize this unity, to understand that we are not alone in the universe, not isolated beings moving through a world that exists outside of us. We are part of the universe, part of the same cosmic dance that moves through the stars and the galaxies, the planets and the nebulae. The self is not confined to the body, not limited by the physical world, but is a field of awareness that stretches across the dimensions, touching everything, connecting everything in a web of relation that transcends time and space.

This understanding brings with it a profound sense of responsibility, for if we are not separate from the universe, then the actions we take, the choices we make, affect not only ourselves but the entire cosmos. Every thought, every action, every moment of awareness sends ripples through the fabric of existence, shaping the reality we inhabit, shaping the reality of those around us. The self is not an isolated point of consciousness, but a node in the cosmic web, a part of the larger whole that is constantly interacting with the field of existence.

In this light, love, compassion, and empathy are not mere emotions, not mere reactions to the world around us. They are reflections of the deeper truth of unity, expressions of the interconnectedness that binds all things together in a web of relation. To love, in this sense, is to recognize the oneness of existence, to see the self not as separate from the other, but as part of the same being, the same consciousness that moves through all things. Compassion becomes an act of recognition, an acknowledgment of the unity that underlies the surface of reality, a recognition that we are all part of the same flow, the same dance of existence.

But this unity is not confined to the human realm. It extends to all things, to all beings, to the stars and the planets, to the trees and the rivers, to the animals and the stones. The universe is not a collection of separate objects, not a machine that operates independently of consciousness. It is a living, breathing entity, a manifestation of the same consciousness that moves through us, that shapes the reality we experience. The stars are not distant, cold objects, but manifestations of the same energy, the same consciousness that flows through our veins, that shapes our thoughts and our dreams.

In this vision of unity, the boundaries between self and universe dissolve, revealing a truth that is at once beautiful and humbling: that we are not separate from the universe, not isolated beings moving through a world that exists outside of us. We are the universe, part of its infinite unfolding, part of its eternal dance. The self, far from being a fixed point of consciousness, is a dynamic process, a wave of awareness that moves through the field of existence, shaping and being shaped by the universe.

This understanding brings with it a profound sense of peace, a recognition that the struggles and the sorrows of life are not the result of separation, but of the illusion of separation. The pain we feel, the suffering we endure, arises from the belief that we are alone, that we are isolated beings moving through a world that is indifferent to our existence. But this belief is an illusion, a veil that hides the deeper truth of unity, the truth that we are part of the same flow, the same dance, the same unfolding of existence.

To see through this illusion is to awaken to a new way of being, a way that embraces the unity of existence, that moves through the world with a sense of connection, a sense of oneness with all things. It is to recognize that the self is not separate from the universe, not isolated from the world around it, but is part of the same consciousness that moves through the stars and the galaxies, the planets and the nebulae. The self is not confined to the body, not limited by the physical world, but is a field of awareness that stretches across the dimensions, touching everything, connecting everything in a web of relation that transcends time and space.

In the quiet moments of reflection, we may catch a glimpse of this deeper reality, a flicker of the infinite presence that lies within and around us. We may come to understand that separation is an illusion, a veil that hides the deeper truth of unity, the truth that we are not alone in the universe, but are part of the same cosmic web, part of the same field of existence. And in this understanding, we may find a new way of being, a new way of living, a new way of understanding ourselves and the world around us—a way that embraces the unity of existence, that moves through the world with a sense of grace, a sense of connection, shaping the course of reality with every thought, every action, every breath.

CHAPTER 16: THE SILENT SYMPHONY OF ENERGY AND FORM

Beneath the visible world of form, there lies an eternal symphony—silent, yet ever-present—an unseen rhythm that pulses through the veins of existence. It is a harmony that weaves the stars into their orbits and whispers through the atoms that compose our flesh. This symphony is the essence of energy, the primal force that animates all things, unseen yet omnipresent, silent yet infinitely powerful. To grasp its nature is to understand that matter, in all its solidity, is but a fleeting expression of something far deeper, far more elusive—a vibration, a movement, a dance of forces that transcend what the eyes perceive and the hands can touch.

Energy is the true foundation of reality, a force that permeates everything, uniting the vastness of the cosmos with the intimacy of the soul. In its purest form, it is formless, shapeless, boundless—a sea of potential, waiting to be shaped by consciousness, waiting to be crystallized into the forms we know as matter. But energy does not merely exist; it moves, flows, dances through the universe, weaving together the worlds of the seen and the unseen, the physical and the spiritual, in a seamless continuum of becoming.

What we perceive as solid, as tangible, as real, is but the surface of this deeper ocean. The chair upon which we sit, the earth beneath our feet, the stars that burn in distant galaxies—these are not solid things, fixed in place and time. They are vibrations, patterns of energy that have momentarily coalesced into form, fleeting manifestations of a deeper reality that is ever-changing, ever-shifting. Matter is not the foundation of existence, but its surface—a thin veil of solidity that conceals the ocean of energy that lies beneath.

This understanding of reality challenges the way we perceive the world, for it suggests that the material realm, far from being the most real or fundamental, is a transient phenomenon, a ripple on the surface of an endless sea. The true nature of existence is not found in the solidity of objects, but in the invisible forces that move through them, in the energy that gives them life and form. To perceive the world in this way is to awaken to a new vision of reality, one in which energy, not matter, is the true essence of existence.

In the quantum world, this truth becomes apparent. At the smallest scales, particles do not behave as solid objects, but as waves of probability, as fields of potential that can exist in multiple states at once, depending on how they are observed. They flicker in and out of existence, momentarily crystallizing into form before dissolving back into the ocean of energy from which they arose. In this view, reality is not fixed or stable, but a fluid dance of energy and form, a continuous process of becoming in which nothing is ever truly static.

But what is the relationship between energy and form? How does the formless become form, and what role do we, as conscious beings, play in this process? The answer lies in the nature of consciousness itself, for it is consciousness that shapes the formless into form, that gives structure and meaning to the swirling sea of potential. Consciousness is not a passive observer of reality, but an active participant in its creation. It is the force that collapses the waves of potential into the particles of matter, the force that brings the silent symphony of energy into the visible world of form.

In this sense, we are not merely inhabitants of the universe, but creators of it. Every thought, every action, every moment of awareness shapes the energy that surrounds us, gives it form, brings it into being. The universe is not something that happens to us; it is something we help to create, moment by moment, with every breath we take. We are not separate from the energy that animates the cosmos; we are part of it, woven into the same symphony that moves through the stars and the galaxies, through the mountains and the seas.

This understanding brings with it a profound sense of responsibility, for if we are creators of reality, then the world we experience reflects our consciousness, a manifestation of the energy we bring into it. The forms we encounter, the experiences we have, are not fixed or predetermined, but are shaped by the vibrations of our thoughts, our emotions, our awareness. The universe is a mirror, reflecting back to us the energy we project into it, shaping itself according to the patterns of our consciousness.

But this creation is not a solitary act, for we are not alone in this symphony. The energy that moves through us is the same energy that moves through all things, through all beings, through the entire cosmos. We are not isolated creators, shaping a world that exists apart from us. We are part of a collective process, a shared dance of energy and form in which all things participate. The vibrations of our thoughts, our emotions, ripple outward through the field of existence, touching everything, affecting everything, shaping not only our own reality but the reality of others as well.

In this sense, energy is not merely a force that moves through the physical world, but a force that moves through the realms of thought, of emotion, of spirit. It is the essence of all things, the thread that connects the visible and the invisible, the material and the immaterial. It flows through the stars and the galaxies, through the earth and the oceans, through the minds and hearts of all beings. It is the silent symphony that unites us all, that weaves us together in a web of connection that transcends the boundaries of time and space.

To live with this awareness is to understand that we are not separate from the world around us, not separate from the energy that animates the cosmos. We are part of it, part of the same symphony, the same dance of creation and destruction, of birth and rebirth. The energy that moves through the stars is the same energy that moves through our bodies, that shapes our thoughts, our dreams, our desires. We are not passive observers of the universe, but active participants in its unfolding, co-creators of the reality we experience.

This vision of reality brings with it a new understanding of the self, for if we are not separate from the energy that animates the cosmos, then the self, too, is not a fixed or isolated entity, but a dynamic process, a field of energy that is constantly interacting with the world around it. The self is not a thing, not a solid object that exists independently of the universe, but a wave of consciousness, a vibration that moves through the field of existence, shaping and being shaped by the energy that surrounds it.

In this light, identity becomes fluid, malleable, a process of becoming rather than a fixed state of being. The self is not a static thing that remains unchanged throughout life, but a dynamic presence that is constantly evolving, constantly interacting with the energy of the universe. Just as the particles of the quantum world are not fixed, but exist in a state of constant flux, so too is the self a process of becoming, a wave of consciousness that moves through the dimensions of reality, shaping and being shaped by the silent symphony of energy and form.

But if the self is a wave of consciousness, if identity is fluid and ever-changing, then what does this mean for our understanding of life, of existence? It means that life is not a linear journey from birth to death, but a dance of energy and form, a continuous process of becoming in which nothing is ever truly fixed or final. The forms we take, the experiences we have, are not predetermined or permanent, but are shaped by the energy that moves through us, by the consciousness that interacts with the quantum field.

In this dance, we are not merely actors playing out a script that has already been written. We are co-creators, shaping the story as it unfolds, shaping the forms that arise from the ocean of energy that surrounds us. The universe is not a fixed or static thing, but a living, breathing entity, a manifestation of consciousness that is constantly evolving, constantly being shaped by the awareness that moves through it.

But what of death? If life is a dance of energy and form, if the self is a wave of consciousness that moves through the quantum field, then death is not an ending, not a cessation of being, but a transformation, a shift from one state of energy to another. The form may dissolve, the body may return to the earth, but the energy that animates the self, the consciousness that shapes reality, does not disappear. It continues to move through the universe, continues to participate in the silent symphony of creation and destruction, of birth and rebirth.

In this vision, death is not something to be feared, not something to be resisted, but something to be understood as part of the larger process of becoming. It is a transition, a moment in the dance of energy and form, a shift from one state of being to another. The self, in its essence, is not lost in death, for the self is not the body, not the form that it temporarily inhabits. The self is the energy that moves through the form, the consciousness that shapes reality, the wave of awareness that continues to dance through the cosmos, even as the form dissolves back into the ocean of potential.

To live with this awareness is to embrace the fluidity of existence, to understand that life is not a linear journey, but a process of becoming, a dance of energy and form that never truly ends. It is to recognize that the forms we take, the experiences we have, are not fixed or permanent, but are shaped by the energy that moves through us, by the consciousness that interacts with the quantum field. It is to understand that we are not separate from the universe, but part of its endless unfolding, part of the same symphony, the same dance of creation and destruction, of birth and rebirth.

In the quiet moments of reflection, we may catch a glimpse of this deeper reality, a flicker of the silent symphony that moves through us and through the cosmos. We may come to understand that we are not passive observers of the universe, but active participants in its unfolding, co-creators of the reality we experience. And in this understanding, we may find a new way of being, a new way of living, a new way of understanding ourselves and the world around us—a way that embraces the dance of energy and form, that moves through the world with grace and awareness, shaping the course of existence with every thought, every action, every breath.

CHAPTER 17: THE INFINITE MIRROR

Consciousness And The Reflective Universe

There is a mirror suspended within the heart of existence—a mirror so vast that it reflects not just the visible world but the very soul of creation itself. This mirror is not of glass, nor of any material substance, but of consciousness, an invisible lens through which reality gazes upon itself. In its depths, everything is mirrored: the stars spinning in their cosmic dance, the oceans surging against the shore, the minds of every sentient being shaping the world through thought and perception. To live is to stand before this infinite mirror, to be both the seer and the seen, the observer and the observed, as the universe reflects back the awareness that flows through it.

But what is consciousness if not the light that illuminates this mirror? It is the flame that casts the shadows into form, the essence that animates the motion of the stars and stirs the silent pulse of life. It is through consciousness that the universe knows itself, for the cosmos is not an indifferent stage, a mechanistic structure unfolding in blind obedience to physical laws. Rather, it is a living, breathing entity—an entity that perceives, that reflects, that interacts with the awareness that flows through every corner of its vastness.

In this reflective universe, the boundaries between self and world dissolve. We are not separate from the reality we experience; we are co-creators, shaping the reflections that come back to us with every thought, every feeling, every action. The universe is not a passive canvas upon which life is painted; it is a mirror, constantly reflecting the energy of consciousness, the light of awareness, the vibrations of thought. Every ripple in the mind sends waves across this mirror, distorting or clarifying the reflection that meets us at every turn.

This understanding—that reality reflects consciousness—challenges the deeply ingrained belief that the world is something external, something that happens to us rather than something we participate in. But once this veil is lifted, we begin to see that what appears before us is not separate from the mind that perceives it. The landscape of reality is sculpted by the forces of perception, and those forces are shaped by the deeper currents of awareness, by the energy of consciousness that moves through all things.

To contemplate this is to recognize that we are not passive witnesses to the universe, but active participants in its unfolding. We stand before the mirror of existence and see ourselves reflected in the cosmos, for the universe itself reflects the consciousness that perceives it. The stars that burn in distant galaxies, the trees that stretch their roots deep into the earth, the rivers that carve their paths through ancient stones—all are reflections of the same awareness, the same light of being that moves through us.

But what does it mean for the universe to reflect consciousness? It means that the reality we experience is not fixed or static, not something that exists independently of the mind that perceives it. Reality is a process, a fluid and dynamic unfolding that is shaped by the energy of awareness, by the vibrations of thought and intention. Just as the quantum world reveals that particles exist in a state of potential until they are observed, so too does the universe exist in a state of infinite possibility, waiting to be shaped by the consciousness that perceives it.

This reflective nature of reality is not a metaphor but a fundamental aspect of existence. Every thought, every feeling, every moment of awareness sends ripples through the fabric of the universe, shaping the reflection that comes back to us. The world we experience is a mirror of our consciousness, a reflection of the energy we project into it. To change the reflection is not to change the world itself, but to change the consciousness that perceives it, to shift the energy that moves through the mirror of reality.

In this sense, the universe is not something that happens to us, but something that happens through us. We are the light that illuminates the mirror, the force that shapes the reflection. The stars are not distant and cold but are part of the same consciousness that burns within us. The mountains and rivers, the forests and skies—

all are expressions of the same awareness, the same life force that moves through every cell of our being. The separation we perceive between self and world, between mind and matter, is an illusion born of limited perception, a distortion in the infinite mirror that hides the deeper truth of unity.

But if the universe reflects consciousness, then what role do we, as conscious beings, play in this cosmic dance of reflection? We are both the creators and the creations, the shapers and the shaped. The thoughts we think, the emotions we feel, the actions we take—these are not isolated events, but ripples in the infinite mirror of existence, waves of consciousness that shape the reality we experience. The universe responds to the energy we project into it, reflecting back to us the vibrations of our awareness, the light of our consciousness.

This understanding brings with it a profound sense of responsibility, for if reality reflects consciousness, then the world we experience is a mirror of the energy we bring into it. The reflections we encounter are not fixed or predetermined, but are shaped by the thoughts we think, the emotions we feel, the awareness we bring to each moment. The universe is not a passive stage, but an active participant in the creation of reality, a mirror that reflects back to us the energy we project into it.

In this reflective universe, every moment is an opportunity to shape reality, to shift the reflection that comes back to us. The energy we project into the world —the thoughts, the emotions, the intentions—shape the reflections we encounter, the experiences we have. To change the reflection is not to change the world itself, but to change the consciousness that perceives it, to shift the energy that moves through the mirror of existence.

But what does it mean to shift consciousness, to change the reflection? It means to awaken to the deeper truth of unity, to recognize that the separation we perceive between self and world, between mind and matter, is an illusion. The self is not an isolated point of consciousness, not a lone traveller moving through a world that exists outside of it. The self is a field of awareness, a wave of consciousness that moves through the universe, shaping and being shaped by the reflections that come back to it.

This awakening to unity is not a passive realization, not a distant, abstract concept. It is an active process, a continuous unfolding of awareness that shapes the way we experience reality. To awaken to unity is to see the universe not as a collection of separate objects, but as a living, breathing entity, a manifestation of consciousness that is constantly interacting with the awareness that flows through it. The stars, the mountains, the rivers—these are not separate from the self, but are reflections of the same consciousness, the same awareness that moves through all things.

To live with this awareness is to move through the world with a sense of grace, a sense of connection to the greater whole. It is to recognize that every moment is an opportunity to shape reality, to create the reflections we wish to see in the mirror of existence. The thoughts we think, the emotions we feel, the actions we take—these are not isolated events, but waves of consciousness that shape the reality we experience. The universe is a mirror, reflecting back to us the energy we project into it, and the reflections we encounter are a manifestation of the consciousness that moves through us.

In this reflective universe, love, compassion, and empathy are not mere emotions, not mere reactions to the world around us. They are reflections of the deeper truth of unity, expressions of the interconnectedness that binds all things together in a web of relation. To love, in this sense, is to see the self reflected in the other, to recognize that the separation we perceive between self and world, between self and other, is an illusion. Compassion becomes an act of recognition, an acknowledgment of the unity that underlies the surface of reality, a recognition that we are all part of the same consciousness, the same awareness that moves through the universe.

This understanding brings with it a profound sense of peace, for if reality reflects consciousness, then the world we experience is a mirror of the energy we project into it. The reflections we encounter are not fixed or predetermined, but are shaped by the thoughts we think, the emotions we feel, the awareness we bring to each moment. The universe is a mirror, reflecting back to us the energy we project into it, and the reflections we encounter are a manifestation of the consciousness that moves through us.

To live with this awareness is to embrace the reflective nature of reality, to see the world as a mirror of the consciousness that perceives it. It is to recognize that the separation we perceive between self and world, between mind and matter, is an illusion, a veil that hides the deeper truth of unity. The self is not an isolated point of consciousness, but a wave of awareness that moves through the universe, shaping and being shaped by the reflections that come back to it.

In the quiet moments of reflection, we may catch a glimpse of this deeper reality, a flicker of the infinite mirror that reflects the universe back to itself. We may come to understand that the world we experience is not separate from the consciousness that perceives it but reflects the energy we project into it. And in this understanding, we may find a new way of being, a new way of living, a new way of understanding ourselves and the world around us—a way that embraces the reflective nature of reality, that moves through the world with grace and awareness, shaping the course of existence with every thought, every action, every breath.

CHAPTER 18: THE BREATH OF CREATION

Sound, Vibration, And The Birth Of Form

In the beginning, before light touched the newborn stars and before matter coalesced into shape, there was vibration. It hummed through the emptiness, a silent song of potential, a rhythm that pulsed with the infinite possibilities of existence. This vibration, unseen and unheard, was the first breath of creation—the primordial sound that gave rise to form, to life, to consciousness. It is a vibration that continues to ripple through the fabric of reality, shaping the world we see and the worlds that remain hidden from our view.

The universe, at its core, is not composed of static particles, of inert matter suspended in the void. It is a living, breathing symphony of vibrations, a field of energy that pulses with the rhythm of existence itself.

Every atom, every molecule, every star and planet is a manifestation of this cosmic rhythm, a note in the great song of creation. To understand the nature of reality is to understand that everything, at its most fundamental level, is vibration—a vibration that moves through the silent depths of space, shaping matter and energy into the forms we perceive.

Sound, in this context, is not merely a sensory experience, not merely the vibrations that reach our ears and are interpreted by our minds. Sound is the essence of creation, the force that brings the formless into form, the invisible wave that moves through the quantum field, shaping reality with every pulse. It is said that the universe was spoken into existence, that the first act of creation was a word, a sound that reverberated through the void, bringing light and life into being. This is not just a poetic metaphor but a reflection of a deeper truth—a truth that reveals the power of sound and vibration to shape the very fabric of reality.

In the quantum world, this truth becomes apparent. Particles, the building blocks of matter, do not exist as solid, unchanging objects. They are waves of potential, fields of energy that vibrate at different frequencies, only collapsing into form when observed. These vibrations are not random; they are part of the greater rhythm of the universe, a rhythm that moves through everything, shaping the world we experience. The chair upon which we sit, the earth beneath our feet, the stars above—these are not solid things, but vibrations, patterns of energy that have temporarily crystallized into form.

But what of the self? What of consciousness? If the universe is a field of vibration, then the self, too, is a wave, a rhythm, a vibration that moves through the field of existence, shaping and being shaped by the forces that surround it. The self is not a fixed entity, not a solid object that exists independently of the world. It is a process, a dance of vibrations that moves through the dimensions of reality, interacting with the waves of energy that form the fabric of the cosmos. The thoughts we think, the emotions we feel, the experiences we have—these are not separate from the vibrations of the universe but are part of the same rhythm, part of the same song.

To understand this is to recognize that the self is not something that exists apart from the world, but is part of the world, part of the vibration that moves through all things. The boundaries we perceive between self and world, between mind and matter, are not real. They are illusions, born from the limitations of perception, from the mind's attempt to impose order on the infinite complexity of existence. In reality, there is no separation —only vibration, only rhythm, only the great song of creation that moves through everything, connecting all things in a web of sound and light.

This song is not static; it is constantly evolving, constantly shifting in response to the vibrations of consciousness, to the energy of thought and emotion. The universe is a living, breathing entity, a field of energy that is shaped by the vibrations that move through it. And we, as conscious beings, are part of this process, part of this unfolding. The thoughts we think, the emotions we feel, the words we speak—these are not isolated events, not random occurrences, but vibrations that ripple outward through the fabric of existence, shaping the reality we experience.

Sound, in this sense, is not just an auditory phenomenon but a creative force. It is the breath of creation, the vibration that brings potential into form, that shapes the waves of energy into the patterns of matter and life. The words we speak, the sounds we make, are not just expressions of thought or feeling but acts of creation, waves of vibration that interact with the quantum field, shaping reality with every pulse. To speak is to create, to bring into being the vibrations that will shape the world around us.

But this creation is not limited to the sounds we make. Every thought, every emotion, every intention is a vibration, a wave of energy that moves through the field of existence, shaping reality with every pulse. The universe responds to these vibrations, reflecting back to us the energy we project into it, the rhythm of our consciousness. The reality we experience is not separate from the vibrations we create; it reflects them, a mirror of the energy we bring into the world.

In this reflective universe, every moment is an opportunity to shape reality, to create the vibrations that will bring the formless into form. The energy we project into the world—the thoughts we think, the emotions we feel, the intentions we hold—shapes the vibrations that move through the quantum field, shaping the reality we experience. The universe is not a fixed or static thing, but a living, breathing entity, a manifestation of the vibrations that move through it.

This understanding brings with it a profound sense of responsibility, for if the universe is shaped by vibration, then the world we experience reflects the energy we bring into it. The thoughts we think, the emotions we feel, the words we speak—these are not isolated events, but vibrations that ripple outward through the fabric of existence, shaping the reality we encounter. The universe is a mirror, reflecting back to us the vibrations we project into it, the rhythm of our consciousness.

But what of silence? If sound is the breath of creation, if vibration is the essence of reality, then what role does silence play in this cosmic symphony? Silence, in this context, is not the absence of sound but the space from which sound arises, the potential from which vibration is born. It is the void from which the universe was spoken into being, the stillness that precedes creation, the quiet that holds within it the infinite possibilities of existence. Silence is not emptiness; it is fullness, the pregnant pause before the first note is struck, the space that holds within it the potential for all things.

To sit in silence is not to retreat from the world, but to enter into the heart of creation, to touch the stillness from which all things arise. It is to listen to the silent song of the universe, to attune oneself to the deeper vibrations that move through the fabric of reality. In this silence, we find the space to create, the space to shape the vibrations that will bring the formless into form. The mind, in its quiet state, becomes a mirror, reflecting the vibrations of the universe, shaping the energy that moves through it.

In this sense, meditation is not a withdrawal from the world but an act of creation, a process of attuning oneself to the vibrations of the universe, of shaping the energy that moves through the mind. To meditate is to listen to the silent symphony of existence, to become aware of the vibrations that shape reality, to consciously shape the energy that moves through the self. It is a way of participating in the creation of reality, of shaping the vibrations that will bring the formless into form.

This understanding of sound, of vibration, of silence brings with it a new way of seeing the world, a new way of experiencing reality. The world is not a fixed or static thing, but a living, breathing entity, a manifestation of the vibrations that move through it. The self, too, is not a fixed or isolated entity but a wave of consciousness, a vibration that moves through the field of existence, shaping and being shaped by the rhythms of the universe.

To live with this awareness is to recognize that every moment is an opportunity to create, to shape the vibrations that will bring the formless into form. The thoughts we think, the emotions we feel, the words we speak—these are not isolated events, but vibrations that ripple outward through the fabric of existence, shaping the reality we encounter. The universe is a mirror, reflecting back to us the vibrations we project into it, the rhythm of our consciousness.

In the quiet moments of reflection, we may catch a glimpse of this deeper reality, a flicker of the silent symphony that moves through us and through the cosmos. We may come to understand that sound is not merely a sensory experience but the breath of creation, the vibration that shapes the world we see and the worlds that remain hidden from our view. And in this understanding, we may find a new way of being, a new way of living, a new way of understanding ourselves and the world around us—a way that embraces the creative power of sound, the infinite potential of vibration, and the silent space from which all things arise.

The universe is not separate from us. It is alive with the rhythms of our being, and through this living, breathing symphony, we are both shaped and shapers, both the instrument and the player, both the silence and the sound.

CHAPTER 19: THE ALCHEMY OF THOUGHT

Manifestation And The Quantum Mind

There is a subtle magic woven into the fabric of thought, a power that whispers through the corridors of consciousness, shaping the very essence of reality. Thought, that invisible force born from the depths of the mind, is not merely an idle flicker within the brain. It is an act of creation, an alchemical process that transforms the unseen into the seen, the potential into the tangible. To think is to conjure, to summon forth the latent energies of the universe and shape them into form. Yet, this alchemy is far more intricate than we imagine, for it occurs not in isolation but within the quantum field—a realm where the boundaries between mind and matter dissolve, where thought and reality entwine in a dance of endless becoming.

The ancient philosophers spoke of thought as the seed of reality, a force that brings into existence the world we experience. In the modern age, quantum physics has given new life to this idea, revealing that reality, at its core, is not a fixed or solid thing but a field of potential—a field shaped by the act of observation, by the consciousness that interacts with it. In this quantum field, particles do not exist in any definite state until they are observed; they hover in a state of possibility, a state of pure potential, waiting to be collapsed into form by the act of awareness. This suggests that reality is not something that exists independently of consciousness but is, instead, shaped by it—shaped by the thoughts we think, the intentions we hold, the awareness we bring to each moment.

To understand this is to recognize that thought is not a passive process, not a mere mental activity that takes place within the confines of the brain. Thought is a force, a vibration that moves through the quantum field, shaping reality with every pulse. It is the bridge between the seen and the unseen, the formless and the formed, the potential and the actual. When we think, we do more than process information; we shape the world around us, bringing into being the experiences we encounter. The universe, in this sense, is a mirror, reflecting back to us the vibrations of our consciousness, the energy of our thoughts.

But what is thought, truly? Is it merely a product of the brain, a byproduct of neural activity, or is it something far deeper, far more profound? In the quantum view, thought is not confined to the brain, not limited to the physical processes of the body. It is a field of energy, a wave of consciousness that moves through the quantum field, interacting with the particles of matter, shaping the reality we experience. Thought, in this sense, is not an isolated phenomenon but a part of the larger process of creation, a force that moves through the universe, bringing potential into form.

This understanding of thought as a creative force challenges the conventional view of reality, for it suggests that the world we experience is not something that happens to us but something we help to create. Every thought, every intention, every moment of awareness shapes the quantum field, collapsing the waves of potential into the particles of reality, bringing into being the world we encounter. The universe is not a fixed or static thing but a dynamic process, a field of energy that is constantly being shaped by the consciousness that interacts with it. To think is to create, to participate in the unfolding of reality, to bring into being the experiences that shape our lives.

This process of creation is not random or arbitrary but is governed by the laws of alchemy—the ancient science of transformation that reveals the hidden connections between mind and matter, between thought and form. In alchemy, the philosopher's stone was said to be the key to transmutation, the substance that could turn base metals into gold, transform the mundane into the sublime. But the true philosopher's stone is not a physical substance; it is the mind itself, the consciousness that shapes reality through the alchemical process of thought. To master this process is to become an alchemist of the mind, a creator of reality who understands the power of thought to shape the world.

But how does this alchemy of thought work? How does a simple idea, a fleeting thought, become a tangible experience, a manifested reality? The answer lies in the nature of vibration, for thought, like everything else in the universe, is a vibration, a wave of energy that moves through the quantum field. When we think, we send out vibrations into the universe, vibrations that interact with the field of potential, shaping the energy that surrounds us. These vibrations, in turn, attract to us experiences that resonate with the energy of our thoughts, experiences that reflect the vibrations we project into the world.

This is the law of attraction, the law that governs the alchemy of thought, the process by which like attracts like, by which the energy we project into the universe returns to us in the form of experience. When we think positive thoughts, when we hold intentions of love, abundance, and joy, we send out vibrations that resonate with those qualities, attracting to us experiences that reflect those energies. Conversely, when we think negative thoughts, when we hold intentions of fear, lack, and doubt, we send out vibrations that resonate with those qualities, attracting to us experiences that reflect those energies.

But this process is not as simple as thinking a single thought and waiting for it to manifest. The alchemy of thought is a subtle and intricate process, one that requires more than mere wishful thinking. It requires alignment—alignment between thought, emotion, and action, alignment between the mind and the quantum field. To think a thought is to plant a seed, but that seed must be nurtured with intention, with emotion, with belief. It must be aligned with the deeper currents of consciousness, with the energy of the universe, in order to manifest into form.

This alignment is the key to the alchemy of thought, for thought alone is not enough to shape reality. It must be supported by emotion, by the vibrations of feeling that move through the quantum field, amplifying the energy of thought, bringing it into resonance with the field of potential. Emotion is the fuel that drives the creative process, the force that gives power to thought, turning a simple idea into a tangible experience. When thought and emotion are aligned, when the mind and the heart move in harmony, the vibrations of consciousness become powerful enough to shape reality, to bring the formless into form.

But emotion, too, is not enough. Action is required—action that is aligned with thought and emotion, action that is inspired by the vibrations of consciousness. To think a thought, to feel an emotion, is to plant the seed of creation, but action is the process by which that seed is brought into bloom. It is through action that the vibrations of thought and emotion are anchored into the physical world, through action that the energy of consciousness is crystallized into form.

This process of alignment—of thought, emotion, and action—lies at the heart of the alchemy of thought, the process by which we shape the reality we experience. But it is not a process that takes place in isolation, not a solitary act of creation. The universe, too, plays its part, for the quantum field is not a passive canvas upon which reality is painted. It is an active participant in the creative process, a living, breathing entity that responds to the vibrations of consciousness, shaping itself in response to the energy we project into it.

The universe is a mirror, reflecting back to us the vibrations of our thoughts, our emotions, our actions. It responds to the energy we bring into it, shaping reality in accordance with the vibrations of consciousness. This is the law of cause and effect, the law that governs the alchemy of thought, the process by which the energy we project into the universe returns to us in the form of experience. To change the reality we experience, we must change the energy we project into the universe, change the vibrations of thought, emotion, and action that shape the quantum field.

But this process of creation is not linear, not a simple cause-and-effect chain. It is a dynamic, multidimensional process, a process that takes place within the quantum field, where time and space are fluid, where the boundaries between cause and effect dissolve. In the quantum world, particles can exist in multiple states at once, hovering in a state of potential until they are observed, until the act of consciousness collapses the wave of possibility into a single reality. This suggests that the reality we experience is not fixed or predetermined but is shaped by the consciousness that interacts with it, by the thoughts we think, the emotions we feel, the actions we take.

The alchemy of thought, then, is not a linear process but a quantum process, a process that takes place within the field of potential, where all possibilities exist at once. To think is to choose, to collapse the wave of potential into a single reality, to bring into being the experiences that shape our lives. But this process of creation is not a solitary act, for the universe, too, plays its part, responding to the energy we project into it, shaping itself in accordance with the vibrations of consciousness.

This understanding of the alchemy of thought brings with it a profound sense of empowerment, for it reveals that we are not passive observers of the universe, not victims of circumstance, but active participants in the creation of reality. The thoughts we think, the emotions we feel, the actions we take—these are not isolated events, but vibrations that ripple outward through the quantum field, shaping the reality we experience. The universe is a mirror, reflecting back to us the energy we project into it, the vibrations of consciousness that shape the world.

But this understanding also brings with it a sense of responsibility, for if we are the creators of reality, then the world we experience reflects the energy we bring into it. The suffering we see, the challenges we face, are not random occurrences, not events that happen to us, but reflections of the vibrations we project into the world. To change the reality we experience, we must change the energy we bring into it, align our thoughts, our emotions, our actions with the deeper currents of consciousness, with the vibrations of love, compassion, and joy that resonate with the quantum field.

In the quiet moments of reflection, we may catch a glimpse of this deeper reality, a flicker of the alchemy of thought that shapes the world we see and the worlds that remain hidden from our view. We may come to understand that thought is not a passive process but an act of creation, a force that shapes the quantum field, bringing the formless into form. And in this understanding, we may find a new way of being, a new way of living, a new way of understanding ourselves and the world around us—a way that embraces the alchemy of thought, the infinite potential of consciousness, and the creative power that lies within each of us.

Reality is not something that happens to us; it is something that happens through us, shaped by the thoughts we think, the emotions we feel, the actions we take. We are the alchemists of the quantum field, the creators of reality, and through the alchemy of thought, we shape the world we experience with every breath, every moment, every vibration.

CHAPTER 20: THE LUMINOUS FABRIC OF MEMORY

Time, Space, And The Continuum Of Being

Time is an enigma—an invisible thread that weaves through the fabric of existence, binding past, present, and future in a continuum that both eludes and defines our understanding. We perceive it as a linear river, flowing steadily from one moment to the next, carrying us forward like leaves upon its surface. Yet, beneath this surface, time is not what it seems. It is not a single, unbroken current, but a boundless ocean, filled with ripples and waves, eddies and whirlpools, where moments touch and fold upon one another, where memory and presence intermingle, and where the past is not gone, but ever-present, echoing through the corridors of our consciousness.

To contemplate time is to peer into the depths of

this ocean, to recognize that the division between past, present, and future is an illusion, a veil drawn over the true nature of existence. We think of time as something that happens to us, something outside ourselves, a force that pulls us relentlessly toward an unknown future. Yet, time is not external, not separate from the self. It is woven into the fabric of our being, inseparable from the consciousness that perceives it. We do not merely move through time; time moves through us, shaped by our awareness, by the rhythms of thought, by the flow of memory.

Memory, in this sense, is not a mere echo of the past, not a relic of moments gone by. It is a living, breathing force, a luminous thread that connects us to the totality of our experience, to the infinite moments that have shaped and continue to shape our being. The past is not something that recedes into the distance, disappearing into the mists of forgetfulness. It is alive within us, woven into the very essence of who we are, an ever-present companion that walks with us through every moment of our existence. The self, in this view, is not a linear narrative, but a multidimensional being, a wave of consciousness that stretches across the expanse of time, touching the past, the present, and the future simultaneously.

In the quantum world, time behaves in ways that defy our ordinary understanding. Particles can exist in multiple states at once, moving both forward and backward through time, their trajectories shaped not only by the past but by the future as well. This suggests that time, like space, is a dimension—one that can be bent, stretched, folded upon itself, where the boundaries between moments blur and dissolve. If this is true in the realm of particles, then why should it not also be true in the realm of consciousness? What if time, as we perceive it, is not a fixed river, but a fluid, malleable field—one that we can navigate, shape, and reshape through the power of thought and memory?

Memory, in this context, is not a passive recollection of past events, but a creative force—a force that allows us to move through the continuum of time, to touch the past and bring it into the present, to reshape the future through the act of remembering. To remember is not merely to retrieve information from the depths of the mind; it is to travel, to journey through the landscape of time, to touch the moments that have shaped us and to reshape them through the power of awareness. The past is not fixed; it is a living field of potential, one that can be revisited, reimagined, transformed through the act of memory.

This understanding of memory as a creative force challenges the conventional view of time as a linear progression, where the past is fixed and the future is yet to be written. It suggests that the past, far from being a closed chapter, is an open field—one that we can access, interact with, and transform. The memories we hold are not static records of events gone by; they are fluid, malleable constructs, shaped by the consciousness that holds them. To remember is to create, to bring the past into the present and to reshape it in the light of new awareness.

But what does it mean to reshape the past? How can the past, which we think of as something fixed, something that has already happened, be changed? The answer lies in the nature of time itself, for if time is a continuum, if it is not linear but multidimensional, then the past is not something that lies behind us, something that is lost and unreachable. It is something that exists within us, something that is alive in the present moment, something that we can touch, interact with, and transform through the act of consciousness.

To reshape the past is not to change the events that have occurred, but to change our relationship to them, to reimagine the way those events have shaped us, to transform the energy of those moments through the power of awareness. The self is not a fixed entity, not a static being that moves through time unchanged. It is a process, a wave of consciousness that is constantly interacting with the field of time, constantly being shaped and reshaped by the memories it holds. To remember is to participate in this process of becoming, to engage with the past not as a fixed narrative, but as a field of potential, one that can be transformed through the act of awareness.

In this sense, memory becomes a tool of creation, a way of shaping the self, of shaping reality. The memories we hold, the stories we tell ourselves about the past, are not fixed or immutable; they are fluid, malleable constructs, shaped by the consciousness that holds them. To remember is to engage with the creative power of time, to participate in the unfolding of reality, to bring the past into the present and to reshape it in the light of new awareness. The past, far from being a closed chapter, is an open field of potential, one that we can access, interact with, and transform through the power of thought and memory.

But this process of reshaping the past is not merely an intellectual exercise; it is a deeply emotional and spiritual journey, one that requires us to engage with the full spectrum of our being, to confront the shadows of our history, and to bring them into the light of consciousness. The memories we hold are not just stories we tell ourselves about the past; they are living, breathing energies that shape our present and our future. To engage with these memories, to bring them into awareness, is to engage with the deeper currents of consciousness, to participate in the alchemical process of transformation that lies at the heart of being.

This alchemy of memory is not a solitary act, for we are not isolated beings, moving through time as lone travellers. We are part of a larger whole, part of a web of connection that stretches across the dimensions of time and space, linking us to the past, the present, and the future, to the experiences of others, to the collective consciousness of humanity. The memories we hold are not just our own; they are part of the larger story of existence, part of the collective memory of the universe. To engage with memory is to engage with this larger field of consciousness, to participate in the unfolding of the collective story, to contribute to the shaping of reality.

In this sense, the self is not an isolated point of consciousness, not a lone traveller moving through the field of time. It is a wave of awareness that stretches across the dimensions, touching the past, the present, and the future simultaneously, interacting with the collective memory of the universe, shaping and being shaped by the energy of consciousness that moves through it. The memories we hold, the experiences we have, are not just personal; they are part of the larger story of existence, part of the collective unfolding of reality.

This understanding of memory as a creative force, as a way of interacting with the continuum of time, brings with it a profound sense of responsibility, for the past, the present, and the future are not separate from one another. They are intertwined, part of the same field of consciousness, part of the same process of becoming. The memories we hold, the stories we tell ourselves about the past, shape not only our present but our future as well. To engage with memory is to engage with the process of creation, to participate in the unfolding of reality, to shape the world we experience through the power of thought and awareness.

But this process of creation is not limited to the individual; it is a collective act, a shared journey through the field of time, one that links us to the larger whole, to the collective consciousness of humanity, to the story of existence itself. The memories we hold are not just personal; they are part of the collective memory of the universe, part of the larger story that is unfolding through the dimensions of time and space. To engage with memory is to participate in this collective story, to contribute to the shaping of reality, to become a co-creator of the universe.

In this sense, memory becomes a sacred act, a way of connecting to the deeper currents of consciousness, to the larger story of existence. To remember is to engage with the creative power of time, to participate in the unfolding of reality, to bring the past into the present and to reshape it in the light of new awareness. The past is not something that is lost or unreachable; it is alive within us, woven into the fabric of our being, a luminous thread that connects us to the totality of existence.

This understanding of memory brings with it a new way of seeing the self, a new way of understanding reality. The self is not a fixed or isolated entity but a process of becoming, a wave of consciousness that stretches across the dimensions of time, touching the past, the present, and the future simultaneously. The memories we hold are not just reflections of the past but are living, breathing energies that shape our present and our future. To engage with memory is to engage with the process of creation, to participate in the alchemical process of transformation that lies at the heart of being.

In the quiet moments of reflection, we may catch a glimpse of this deeper reality, a flicker of the luminous fabric of memory that connects us to the totality of existence. We may come to understand that time is not a linear progression but a multidimensional field, one that we can navigate, shape, and reshape through the power of thought and awareness. And in this understanding, we may find a new way of being, a new way of living, a new way of understanding ourselves and the world around us—a way that embraces the creative power of memory, the infinite potential of consciousness, and the timeless continuum of being that links us to the past, the present, and the future.

Reality, then, is not a fixed or static thing, not a linear progression from one moment to the next. It is a living, breathing continuum, a field of potential that is shaped by the consciousness that interacts with it. The past, the present, and the future are not separate from one another but are part of the same field of being, part of the same process of becoming. To remember is to participate in this process, to engage with the creative power of time, to bring the past into the present and to shape the future through the power of thought and awareness.

In this way, we are not just travellers moving through the river of time; we are creators, shapers of reality, participants in the unfolding of the universe. And through the luminous fabric of memory, we are connected to the totality of existence, to the infinite moments that have shaped and continue to shape our being, to the collective story of humanity, and to the timeless continuum of being that stretches across the dimensions of time and space.

CHAPTER 21: THE VEINS OF INFINITY

Space, Matter, And The Boundless Field Of Creation

Space is often mistaken for emptiness—a void, a vast silence stretching endlessly between stars, galaxies, and worlds. It appears as a chasm, a boundless nothingness that yawns between all things, separating them with its immeasurable expanse. Yet, this perception of space as an empty container for matter is an illusion. Space is not the absence of substance, but a living field, a breathing medium through which the forces of existence flow. It is the canvas upon which creation paints its endless forms, the silent stage upon which the dance of matter unfolds. Within its invisible web lies the secret to creation itself, for space is not void but teeming with potential—a potential that pulses beneath the surface of reality, waiting to burst into form.

To contemplate space is to contemplate the infinite. It

is a medium not just of distance but of connection, an unseen web that links all things, no matter how far apart they appear. Space is not passive; it is an active participant in the shaping of reality, a field of energy and possibility that influences everything that moves within it. Every star, every atom, every grain of dust is held in the embrace of space, and through that embrace, each one is linked to every other. Space is not a gap between objects but the very medium through which those objects come into being. Without space, there would be no form, no structure, no existence. The emptiness we perceive is, in truth, a field of infinite possibility, a field that gives rise to the forms we know.

In the quantum realm, space is revealed not as an empty container, but as a field of energy, a fabric that vibrates with potential. It is within this field that particles arise, not from nothing, but from the energetic currents that flow through space itself. Matter, in this sense, is not a solid, independent substance but a manifestation of space's hidden energy—a condensation of the invisible forces that weave through the universe. Particles flicker into existence, not as isolated points, but as waves in the vast ocean of space, ripples in the field of potential that stretches beyond our perception. Space is not separate from matter; it is the very essence from which matter emerges.

But what does it mean for space to be alive with potential? What does it mean for the empty stretches between stars and atoms to hum with energy, with the possibility of form? It means that creation is not confined to a single moment, a singular act that occurred in the distant past. Creation is an ongoing process, a continuous unfolding of form from the formless, of matter from energy, of the visible from the invisible. Space is not a backdrop to this process; it is the source, the womb from which all things arise. The universe is constantly being born, constantly emerging from the hidden potential of space, constantly crystallizing into the forms we perceive.

In this vision of space, the distinction between emptiness and matter dissolves. Matter is not separate from space; it is space condensed, space made visible, space brought into form. Every object, every particle, every living being is a wave in the ocean of space, a ripple in the infinite field of potential that stretches through the cosmos. The mountains, the rivers, the stars—all are not separate from space but are expressions of it, manifestations of its hidden energy. To touch matter is to touch space; to see form is to see the invisible forces that give rise to it. There is no gap, no distance, no separation—only the continuous flow of energy from the unseen to the seen, from the formless to the formed.

This understanding of space as the source of creation challenges the conventional view of reality, for it suggests that the world we experience is not a collection of separate objects, not a universe of isolated things scattered across an empty void. The universe, in truth, is a single, interconnected field, a continuous web of energy and form, of space and matter. The separation we perceive between objects, between stars, between beings, is an illusion born of limited perception. In reality, all things are connected, all things are woven together in the fabric of space, all things are expressions of the same underlying field of potential.

Space, then, is not the absence of form but the source of

it. It is the medium through which matter is born, the field from which creation emerges. Every particle, every atom, every star is a wave in the ocean of space, a ripple in the field of energy that stretches through the cosmos. The universe is not a collection of separate things but a continuous flow of energy and form, a seamless web of being that extends through the infinite depths of space.

But if space is the source of matter, if space is alive with potential, then what does this mean for the nature of creation? It means that creation is not a single act, not a moment in time that has passed. Creation is an ongoing process, a continuous unfolding of form from the formless, a continuous birth of matter from the invisible energy of space. The universe is constantly being born, constantly emerging from the hidden potential of space, constantly crystallizing into the forms we perceive. Every moment is a moment of creation, a moment in which the energy of space gives birth to the forms that populate the universe.

This process of creation is not confined to the physical realm, not limited to the birth of stars and galaxies, of atoms and molecules. It extends to the realms of thought, of emotion, of consciousness. Just as space gives rise to matter, so too does it give rise to the forms of the mind, to the thoughts, the ideas, the dreams that shape our experience. The mind, like the body, is a wave in the ocean of space, a ripple in the field of energy that stretches through the cosmos. The thoughts we think, the emotions we feel, the dreams we dream—these are not isolated events, not random occurrences, but expressions of the same underlying field of potential, the same space that gives rise to stars and galaxies.

In this sense, space is the source of all creation, both physical and metaphysical. It is the medium through which the universe unfolds, the field from which all things emerge. Matter, thought, emotion, consciousness —all are waves in the ocean of space, ripples in the infinite field of potential that stretches through the cosmos. The universe is not separate from space but is an expression of it, a manifestation of its hidden energy, a crystallization of its potential into form.

This understanding of space brings with it a new way of seeing the world, a new way of experiencing reality. The objects we perceive, the forms we encounter, are not separate from space but are expressions of it, manifestations of its hidden energy. The mountains, the rivers, the stars, the minds of every sentient being—all are waves in the ocean of space, ripples in the field of energy that stretches through the cosmos. The universe is a single, interconnected field, a seamless web of being that extends through the infinite depths of space.

But what does this mean for the self? What does it mean to be a wave in the ocean of space, a ripple in the field of energy that stretches through the cosmos? It means that the self, like the universe, is not a fixed or static thing, not a separate or isolated entity. The self is a process, a wave of consciousness that moves through the field of space, interacting with the energy of the universe, shaping and being shaped by the forces that flow through it. The thoughts we think, the emotions we feel, the experiences we have—these are not isolated events but waves in the ocean of space, ripples in the field of potential that stretches through the cosmos.

In this view, the self is not separate from the world, not isolated from the universe. The self is part of the same field of energy, part of the same web of being that extends through the infinite depths of space. The thoughts we think, the emotions we feel, the experiences we have—these are not confined to the self, not limited to the body or the mind, but are part of the larger flow of energy that moves through the cosmos. The self is not an island, not a fixed point of consciousness, but a wave in the ocean of space, a ripple in the field of potential that stretches through the universe.

To live with this awareness is to recognize that we are not separate from the world, not isolated from the universe. We are part of the same field of energy, part of the same web of being that extends through the infinite depths of space. The thoughts we think, the emotions we feel, the experiences we have—these are not isolated events but waves in the ocean of space, ripples in the field of potential that stretches through the cosmos. The universe is a mirror, reflecting back to us the energy we project into it, the vibrations of our consciousness that shape the world we experience.

In this vision of reality, the distinction between self and world, between matter and space, dissolves. The universe is not a collection of separate things but a continuous flow of energy and form, a seamless web of being that extends through the infinite depths of space. Matter is not separate from space; it is space condensed, space made visible, space brought into form. The self, too, is not separate from the world but is part of the same field of energy, part of the same process of creation that unfolds through the dimensions of space and time.

This understanding brings with it a profound sense of connection, for if the self is not separate from the world, if matter is not separate from space, then all things are linked, all things are part of the same field of energy, part of the same web of being. The universe is not a random collection of objects but a single, interconnected field, a living, breathing entity that is constantly being born, constantly unfolding, constantly creating itself anew. Space is the source of this creation, the medium through which the universe is born, the field from which all things emerge.

In the quiet moments of contemplation, we may catch a glimpse of this deeper reality, a flicker of the infinite field that lies beneath the surface of form, a glimmer of the hidden energy that gives rise to the universe. We may come to understand that space is not empty, not void, but alive with potential, alive with the forces that shape the world we experience. And in this understanding, we may find a new way of being, a new way of living, a new way of understanding ourselves and the world around us —a way that embraces the infinite potential of space, the boundless field of creation, and the endless flow of energy and form that stretches through the cosmos.

CHAPTER 22: THE VEIL OF PERCEPTION

Illusion, Reality, And The Horizon Of Consciousness

Perception is the lens through which we gaze upon the world, a delicate veil that filters the infinite complexities of reality into forms and sensations that our minds can grasp. It shapes the contours of our experience, drawing lines between what is seen and unseen, what is known and unknowable. Yet, like all veils, perception conceals even as it reveals, cloaking the deeper mysteries of existence in the comforting illusions of the familiar. We walk through life believing that what we see, touch, and understand constitutes reality. But the true nature of existence lies not in the surfaces we encounter but in the depths beneath, in the realms beyond perception, where reality vibrates with a richness that our minds can barely fathom.

To perceive is to partake in an illusion, though not in a way that diminishes the beauty or significance of the world. Rather, the illusion is that perception captures the full truth of existence. Our senses, marvellous though they are, grasp only fragments of the whole. Like the thin light of dawn stretching over a vast horizon, they illuminate only a sliver of what lies beyond. The eyes see colours, but not the infinite wavelengths that dance through the air; the ears hear sounds, but not the subtle vibrations that pulse beneath the threshold of silence. Reality, in its true form, is boundless, infinite—an ocean of potential that stretches far beyond the limits of our perception.

We are creatures bound by the limits of our senses, yet we are also creatures of consciousness, and it is through this consciousness that we glimpse the vastness of what lies beyond. Perception may limit, but consciousness expands, offering us a window into the depths of existence that perception alone cannot touch. In the quiet spaces between thoughts, in the moments of stillness where the mind ceases its ceaseless chatter, we can feel the pulse of a deeper reality, the vibration of a universe that exists not outside us, but within.

What we perceive as the external world, the solid and tangible realm of form, is only the outermost layer of reality, a thin veil that conceals the deeper truths that lie beneath. It is as though we are standing at the edge of a vast ocean, able to see only the surface, while beneath the waves lies an entire world teeming with life, with movement, with complexity. The world we experience through our senses is like this surface—beautiful, intricate, full of wonder—but it is only a fragment of the greater whole. Beneath this surface, the true nature of reality moves in currents and tides that we can scarcely imagine.

Yet, there are moments when the veil of perception thins, when we catch a glimpse of the infinite beneath. These moments are rare, fleeting, like the brief opening of a door that reveals a vast and luminous landscape before it swings shut again. They come to us in dreams, in moments of profound insight, in the stillness of meditation, or in the sudden awe of standing beneath a star-filled sky. In these moments, we sense that the world is not as solid, not as fixed, as it seems. We sense the fluidity of existence, the way reality shimmers and shifts, the way form dissolves into energy and back again, the way the boundaries between self and world, between here and there, dissolve into the infinite.

In the quantum world, this fluidity is revealed in all its complexity. Particles, the very building blocks of matter, do not exist as fixed objects but as waves of potential, hovering in a state of possibility until they are observed, until consciousness collapses their infinite potential into a single reality. This suggests that reality, far from being a fixed and immutable thing, is fluid, malleable, shaped by the act of perception, by the consciousness that interacts with it. The world we see is not the world as it is, but the world as it is shaped by the filters of our perception, by the limitations of our senses, by the horizons of our consciousness.

What does it mean, then, to live in a world where reality is not fixed but fluid, where perception shapes the very fabric of existence, where the boundaries between illusion and truth are constantly shifting? It means that reality is not something that exists outside of us, something that happens to us. Reality is something we participate in, something we help to create. Every thought, every perception, every moment of awareness shapes the world we experience, brings into being the forms and sensations that populate our reality. We are not passive observers of the universe; we are co-creators, shaping reality with every breath, every thought, every glance.

But this creation is not a solitary act, for we are not isolated beings, moving through a world that is separate from us. We are part of the same field of consciousness, the same web of existence that stretches through the universe. The reality we experience is shaped not only by our individual perceptions but by the collective consciousness of all beings, by the interconnected web of awareness that links us to the stars, to the earth, to each other. The world we see reflects this collective consciousness, a mirror that reflects back to us the thoughts, the emotions, the energies that we project into it.

This understanding brings with it a profound sense of responsibility, for if perception shapes reality, then the world we experience reflects the consciousness we bring to it. The beauty, the wonder, the joy we see in the world reflects the beauty, the wonder, the joy we hold within ourselves. And the suffering, the chaos, the confusion we see in the world reflects the suffering, the chaos, the confusion that exists within the collective consciousness of humanity. To change the world, we must change the way we perceive it, change the consciousness we bring to it. The veil of perception is not a barrier that separates us from reality; it is a lens that we can shape, refine, and expand through the power of awareness.

But how do we expand this lens? How do we move beyond the limitations of perception, beyond the illusions of form, to glimpse the deeper truths that lie beneath? The answer lies in consciousness itself, for it is through consciousness that we touch the infinite, that we move beyond the surface of reality and into its depths. Consciousness is not confined to the body, not limited by the senses. It is a field of awareness that stretches beyond the physical, beyond the visible, into the unseen dimensions of existence. To expand perception is to expand consciousness, to move beyond the filters of the senses and into the deeper currents of awareness that flow through the universe.

This expansion of consciousness is not an intellectual exercise, not a process of acquiring more knowledge or understanding more facts. It is a process of deepening awareness, of quieting the mind, of attuning oneself to the subtle vibrations that move through the field of existence. It is a process of letting go of the need to define, to categorize, to separate, and instead, allowing oneself to merge with the flow of reality, to feel the pulse of existence moving through the self. In this state of expanded awareness, perception becomes fluid, malleable, and the boundaries between self and world, between form and formlessness, dissolve.

In this state of expanded consciousness, we begin to see the world not as a collection of separate objects, but as a living, breathing field of energy and form. We begin to feel the interconnectedness of all things, the way every thought, every emotion, every action ripples outward through the web of existence, touching everything, affecting everything. We begin to understand that reality is not something that is fixed, but something that is constantly being created, constantly being shaped by the consciousness that moves through it.

But this understanding also brings with it a sense of humility, for the more we expand our perception, the more we realize how little we truly know. The universe, in its vastness, in its complexity, is beyond the grasp of any single mind, beyond the limits of any single perception. The deeper we delve into the mysteries of existence, the more we realize that reality is not something that can be fully understood or captured by thought. It is something that must be experienced, something that must be felt, something that must be lived. Reality is not a problem to be solved, but a mystery to be embraced, a dance to be danced, a journey to be walked.

This journey is not a linear path, not a straight line from ignorance to understanding, from illusion to truth. It is a spiral, a continuous unfolding, where each moment of awareness brings us closer to the infinite yet reveals how much further there is to go. Every step we take, every veil we lift, reveals new layers of reality, new dimensions of being, new horizons of consciousness. The journey is endless, for the infinite cannot be reached, only experienced, only touched, only glimpsed through the shifting veils of perception.

In this vision of reality, the self is not a fixed or static thing, not a solid entity that moves through the world unchanged. The self, like reality, is fluid, malleable, constantly being shaped and reshaped by the perceptions it holds, by the consciousness that moves through it. To expand perception is to expand the self, to deepen the connection to the field of existence, to merge with the infinite currents of awareness that flow through the universe. The self is not separate from the world but is part of the same field of energy, part of the same web of consciousness that stretches through the stars, through the earth, through the depths of space and time.

In the quiet moments of reflection, we may catch a glimpse of this deeper reality, a flicker of the infinite that lies beneath the surface of perception, a glimmer of the vastness that stretches beyond the limits of the senses. We may come to understand that perception, while necessary, is only the beginning, only the surface of a much deeper truth. And in this understanding, we may find a new way of being, a new way of living, a new way of experiencing the world—a way that embraces the fluidity of reality, the boundless horizon of consciousness, and the infinite potential that lies beyond the veil of perception.

To live with this awareness is to recognize that reality is not something fixed, something separate from us. It is a field of potential, a field of energy, a field of consciousness that we help to shape with every thought, every perception, every moment of awareness. We are not passive observers of the universe, but active participants in the creation of reality. And through the expansion of consciousness, through the deepening of awareness, we can begin to glimpse the infinite that lies beyond the veil, the boundless field of potential that stretches through the cosmos, waiting to be shaped, waiting to be experienced, waiting to be known.

CHAPTER 23: THE SILENT PULSE OF ETERNITY

Cycles, Rhythms, And The Timeless Flow Of Existence

There is a rhythm that pulses beneath the surface of all things, a silent cadence that weaves through the stars, through the seasons, through the delicate dance of life and death. It is a rhythm older than time itself, older than the first flicker of light that ignited the cosmos, older than the earth beneath our feet. It is the pulse of eternity—the endless flow of creation and dissolution, of birth and rebirth, a cycle without beginning or cessation, where the past and future entwine in an eternal present. To feel this pulse is to understand that life does not move in a straight line, but in circles, spirals, endless returns to the same moment, the same breath, the same silent heartbeat that moves through all things.

We live in a world of cycles—night and day, the waxing and waning of the moon, the turning of the seasons, the rise and fall of empires, the birth and death of stars. These cycles are not random or chaotic; they are the outward expressions of a deeper rhythm, a cosmic order that governs the flow of all things. The universe, in its infinite complexity, is not a random collection of events but a carefully orchestrated dance, a symphony of rhythms that move in harmony with the silent pulse of eternity. To live is to be part of this dance, to move with the rhythm of the stars, to rise and fall with the tides of existence, to be swept up in the endless flow of creation and dissolution.

Yet, for all its beauty, this rhythm often eludes our awareness. We live our lives as though time were a straight line, moving steadily from one moment to the next, carrying us toward some distant horizon. We measure our days in hours and minutes, our lives in years, as though time were something that could be captured, divided, controlled. But time is not a straight line. It is a circle, a spiral, a wheel that turns and turns, returning us again and again to the same moments, the same lessons, the same experiences, each time with a new layer of meaning, a deeper understanding.

To recognize this cyclical nature of time is to awaken to a new way of being, a new way of understanding the flow of existence. It is to see that life does not move toward an end point, a final destination, but moves in cycles, in waves, in spirals that carry us deeper into the heart of existence with every turn. The future is not something that lies ahead of us, waiting to be reached; it is something that is constantly being born, constantly emerging from the same eternal present that holds the past and the future within it. Every moment is a return, a rebirth, a new beginning.

This cyclical nature of time is mirrored in the natural world. The sun rises and sets, the moon waxes and wanes, the seasons turn from spring to summer to autumn to winter, only to return again to spring. The earth moves in cycles, in rhythms that repeat themselves endlessly, and we, as creatures of this earth, are part of those rhythms. Our bodies, too, move in cycles—the beating of the heart, the rhythm of the breath, the flow of blood through the veins, the ebb and flow of energy, of life itself. We are not separate from the cycles of the earth; we are part of them, woven into the same cosmic web of rhythm and flow.

But these outward cycles are merely reflections of a deeper rhythm, a rhythm that moves not just through the earth but through the entire cosmos. The stars, too, are born and die in cycles, in waves of creation and dissolution that stretch across billions of years. Galaxies collide and merge, black holes form and collapse, entire universes expand and contract in rhythms that echo the silent pulse of eternity. The universe itself is a living being, a vast organism that moves in cycles, in rhythms that are far beyond our comprehension, yet are reflected in every aspect of our existence.

To understand this is to recognize that the cycles of life and death, of creation and dissolution, are not random or meaningless but are part of the deeper rhythm of the cosmos. Birth and death are not opposites, not endpoints on a linear journey, but are part of the same flow, the same eternal dance of becoming and unbecoming. To be born is not to begin, and to die is not to cease. Both are part of the same cycle, the same rhythm that pulses through the stars, through the earth, through the veins of all living beings.

This understanding brings with it a profound sense of peace, for it reveals that there is no true ending, no final cessation, no ultimate loss. There is only transformation, only the shifting of forms, the dissolution of the old and the birth of the new. Death is not a final departure but a return, a return to the source, to the silent pulse of eternity that moves through all things. And life, too, is not a singular, unbroken journey from birth to death, but a series of returns, a series of cycles that carry us deeper into the heart of existence with every turn.

In the ancient traditions, this cyclical nature of existence was understood and revered. The Wheel of Life, the Ouroboros, the endless knot—these symbols speak of the eternal cycles of birth, death, and rebirth, of the continuous flow of creation and dissolution that governs the universe. To live in harmony with these cycles is to live in harmony with the universe, to recognize that we are not separate from the flow of existence but are part of it, carried along by the same currents that move the stars, that turn the seasons, that give rise to life and death.

Yet, in our modern world, we have lost touch with this deeper rhythm. We have come to see life as a straight line, as a journey from beginning to end, from birth to death, from past to future. We measure our success by how far we have come, by how much we have achieved, as though life were a race, a competition, a quest for some distant goal. But this view of life as a linear journey is an illusion, a distortion of the true nature of existence. Life is not a race, not a quest for achievement, but a dance, a spiral, a rhythm that moves through us and through all things, carrying us back again and again to the same eternal present, the same silent pulse of eternity.

To live in harmony with this rhythm is to let go of the need to control, to let go of the desire for certainty, for linear progress, for final answers. It is to surrender to the flow of existence, to trust in the cycles of creation and dissolution, to recognize that every moment is a return, a rebirth, a new beginning. It is to see that there is no need to rush, no need to strive for some distant goal, for the journey is not about reaching a destination but about moving with the rhythm, about participating in the dance of creation and dissolution that moves through all things.

But this surrender is not a passive resignation; it is an active engagement with the flow of existence. To live in harmony with the cycles of the universe is to participate fully in the dance, to move with the rhythm, to create and to dissolve, to be born and to die, to rise and to fall with the tides of existence. It is to recognize that every moment is an opportunity for creation, for transformation, for rebirth. The cycles of life and death are not something that happens to us but something we participate in, something we help to create with every breath, every thought, every action.

This understanding brings with it a new way of seeing the self, a new way of understanding identity. The self is not a fixed or static thing, not a singular entity that moves through time unchanged. The self, like the universe, is a process, a rhythm, a cycle of becoming and unbecoming, of creation and dissolution. We are not the same person today as we were yesterday, and we will not be the same person tomorrow. Every moment is a death and a rebirth, a shedding of the old and a becoming of the new. The self is not a thing but a process, a wave in the ocean of existence, a ripple in the silent pulse of eternity.

In this view, identity is not something fixed or permanent but something fluid, something that is constantly being shaped and reshaped by the rhythms of existence. The self is a wave that rises and falls, that emerges from the ocean of being and dissolves back into it, only to rise again in a new form, a new moment of awareness. To live with this understanding is to embrace the fluidity of existence, to recognize that we are not separate from the flow of life but are part of it, carried along by the same currents that move the stars, that turn the seasons, that give rise to life and death.

But this fluidity of identity does not mean that the self is meaningless or without substance. On the contrary, it reveals the profound richness and depth of being, the infinite potential that lies within each moment of awareness. To be a wave in the ocean of existence is to be part of the infinite, part of the eternal flow of creation and dissolution that moves through all things. The self is not a separate entity but a moment of awareness, a moment of creation, a moment of participation in the eternal dance of being.

In the quiet moments of reflection, we may catch a glimpse of this deeper reality, a flicker of the silent pulse of eternity that moves through us and through the cosmos. We may come to understand that life is not a straight line but a circle, a spiral, a rhythm that carries us deeper into the heart of existence with every turn. And in this understanding, we may find a new way of being, a new way of living, a new way of understanding.

CHAPTER 24: THE ALCHEMY OF LIGHT AND SHADOW

Duality, Unity, And The Dance Of Opposites

Beneath the visible world, there lies a primordial dance—an eternal interplay of light and shadow, of presence and absence, of creation and dissolution. This dance is not a battle, not a contest between opposites, but a cosmic rhythm that weaves through the very heart of existence. It is an alchemy that transcends division, for within the embrace of light and shadow lies a deeper truth, a unity that pulses through all things. To live is to participate in this dance, to move between the extremes of being, to feel the push and pull of opposites, and to glimpse the deeper unity that holds them together.

In the natural world, this interplay of opposites is ever-present. Day gives way to night, and night returns to day. Summer's warmth is chased by winter's cold, only for spring to rise once more from the frozen earth. Life and death, growth and decay, rise and fall—they are not separate forces, but two sides of the same coin, two expressions of the same eternal cycle. Each exists only because of the other, for without shadow, light would be meaningless; without death, life would have no context. They are bound together, inseparable, locked in an eternal embrace, a dance that shapes the rhythm of the cosmos.

Yet, for all its beauty, the duality of light and shadow is often misunderstood. We live in a world that values light over shadow, that exalts presence over absence, growth over decay. We seek to banish the shadow, to escape the darkness, to hold onto the light as though it were something that could be possessed. But to embrace only one side of this duality is to deny the wholeness of existence, to miss the deeper truth that light and shadow are not adversaries, but partners in the dance of creation. They are the yin and yang, the alpha and omega, the pulse of life itself.

Light is the force of creation, the energy that brings form into being, that illuminates the world and reveals the beauty of existence. It is the spark of consciousness, the flame of awareness that burns within each of us, that allows us to see, to know, to understand. But light, for all its brilliance, is not complete without its counterpart, for shadow is the space from which light emerges, the fertile darkness from which all things are born. It is the void, the potential, the mystery that holds the seeds of creation within it. Without shadow, there would be no place for light to shine, no context in which its brilliance could be known.

Shadow, too, is often misunderstood. It is seen as the absence of light, as something to be feared, something to be avoided. Yet shadow is not the enemy of light, but its complement, its partner in the dance of existence. Shadow is the space of possibility, the field of potential that holds within it the unseen, the unknown, the unformed. It is the womb of creation, the silent space from which all things emerge, the darkness that nurtures the seed until it is ready to burst into the light. To fear the shadow is to fear the very essence of creation, to deny the mystery that lies at the heart of being.

This dance of light and shadow is not confined to the external world, to the cycles of day and night, of life and death. It moves through us, as well, shaping the contours of our inner world, the landscape of our consciousness. Within each of us, there is light and shadow, presence and absence, creation and dissolution. The light is the part of ourselves that we show to the world, the part that is known, seen, and understood. It is the conscious mind, the awareness that shapes our thoughts, our actions, our experiences. But within us, too, is shadow, the unseen, the unknown, the unformed. It is the unconscious mind, the mystery that lies beneath the surface of awareness, the field of potential that holds the seeds of transformation.

To live fully, to be whole, is to embrace both light and shadow, to move with the rhythm of their dance, to understand that they are not separate, but two aspects of the same being. The light reveals, but the shadow holds the mystery; the light creates, but the shadow nurtures. To deny one is to deny the other, for they are bound together, inseparable, locked in an eternal embrace. The self, too, is not a fixed entity, not a singular expression of light, but a dynamic process, a wave of consciousness that moves between light and shadow, between presence and absence, between the known and the unknown.

This understanding of duality as a dance, as an alchemy that transcends division, brings with it a profound sense of unity, for it reveals that the opposites we perceive—light and shadow, life and death, creation and dissolution—are not truly opposites at all. They are different expressions of the same underlying reality, different notes in the same cosmic symphony. The universe, in its infinite complexity, is not divided into separate parts, not split into good and evil, light and dark, being and non-being. It is a whole, a single, interconnected field of energy and consciousness, in which all things are part of the same eternal dance.

But to see this unity, to understand the deeper truth that lies beneath the surface of duality, requires a shift in perception, a widening of awareness. It requires us to move beyond the narrow confines of the mind, beyond the rigid categories of thought, and into the fluid realm of consciousness, where the boundaries between opposites dissolve, where light and shadow merge into one. It requires us to embrace the mystery, to step into the unknown, to surrender to the flow of existence and to trust in the deeper rhythm that moves through all things.

This surrender is not a giving up, not a resignation to fate. It is an active engagement with the alchemy of existence, a conscious participation in the dance of light and shadow, a willingness to move with the rhythm of creation and dissolution, to embrace both the light that reveals and the shadow that conceals. It is a recognition that we are not separate from this dance, not outside observers, but active participants, co-creators of reality. The light and shadow that move through the universe move through us, as well, shaping our thoughts, our actions, our experiences. To live fully is to embrace this movement, to dance with the rhythm of being, to create and to dissolve, to rise and to fall with the tides of existence.

This understanding brings with it a new way of seeing the self, a new way of understanding identity. The self is not a fixed or static thing, not a singular entity that moves through time unchanged. The self is a dynamic process, a wave of consciousness that moves between light and shadow, between presence and absence, between the known and the unknown. To be fully alive is to embrace this fluidity, to recognize that we are not defined by the light alone, but by the shadow as well, by the mystery that lies beneath the surface of awareness, by the potential that is held within the darkness.

In the ancient traditions, this dance of light and shadow was understood and revered. The alchemists spoke of the marriage of opposites, of the union of sun and moon, of the transmutation of base metal into gold. But this alchemy was not merely a physical process; it was a spiritual one, a recognition that the path to wholeness lies not in the rejection of one side of the duality, but in the integration of both, in the embrace of light and shadow, of creation and dissolution, of being and non-being. The goal of the alchemist was not to escape the darkness, but to transform it, to bring the light into the shadow and to reveal the hidden gold that lies within.

This process of alchemy, this marriage of opposites, is not confined to the ancient traditions. It is a universal process, a cosmic rhythm that moves through all things, through the stars, through the earth, through the hearts and minds of all beings. It is the process by which the universe creates itself, by which light and shadow come together to give birth to form, by which the formless is made manifest. To live with this awareness is to participate in this process, to become an alchemist of the soul, a creator of reality who understands the power of light and shadow, the power of the dance of opposites, the power of the alchemy of existence.

This understanding of duality as an alchemical process, as a dance of creation and dissolution, brings with it a profound sense of peace, for it reveals that there is no true division, no ultimate separation, no final conflict between light and shadow, between being and non-being. There is only the dance, only the flow of existence, the eternal rhythm of becoming and unbecoming, of creation and dissolution. The opposites we perceive are not separate forces, but different expressions of the same underlying reality, different aspects of the same being. To embrace both is to embrace the whole, to see the unity that lies beneath the surface of duality, to participate fully in the dance of existence.

In the quiet moments of reflection, we may catch a glimpse of this deeper reality, a flicker of the alchemical process that moves through us and through the cosmos. We may come to understand that light and shadow are not adversaries, but partners in the dance of creation, that being and non-being are not separate, but are part of the same eternal flow of existence. And in this understanding, we may find a new way of being, a new way of living, a new way of understanding ourselves and the world around us—a way that embraces the alchemy of light and shadow, the unity of duality, and the eternal dance of opposites that shapes the rhythm of the cosmos.

To live with this awareness is to dance with the rhythm of existence, to move with the flow of creation and dissolution, to embrace both the light that reveals and the shadow that conceals. It is to recognize that we are not separate from the universe, not outside observers of the dance, but active participants, co-creators of reality. And through this dance, through this alchemical process, we come to know the deeper truth of existence, the unity that lies beneath the surface of duality, the silent pulse of eternity that moves through all things.

CHAPTER 25: THE WHISPERING THREAD OF DESTINY

Free Will, Fate, And The Interwoven Path Of Being

Somewhere between the shadowed realms of fate and the luminous possibility of free will, there exists a delicate thread—an unseen filament of destiny that winds its way through the currents of time and space, connecting all things in its silent embrace. This thread, ethereal and fine, moves with both precision and mystery, weaving together the choices of beings and the intricate designs of the universe. To live is to walk along this thread, to feel its subtle pull even as we make our choices, to experience both the freedom and the predestination that flow through the currents of existence like two entwined rivers.

But what is this thing called destiny? Is it a fixed path, etched into the fabric of time before we are even born, or is it something more fluid, more malleable, shaped by the choices we make and the forces we summon through the will? This question has haunted the minds of philosophers and mystics for millennia, for destiny, like the very universe itself, is an enigma—an interplay between forces seen and unseen, known and unknowable, where the boundaries between freedom and fate dissolve into the great ocean of possibility.

To walk the path of life is to exist within this paradox, to dance between the pull of destiny and the power of free will, between the unseen forces that guide us and the choices we make along the way. We often speak of fate as though it were an external force, something that happens to us, something over which we have no control. And yet, there is a deeper truth that whispers through the silence, a truth that suggests that fate and free will are not opposites but two sides of the same coin, two threads in the same cosmic fabric. We are not merely puppets of fate, nor are we masters of our destiny. We are co-creators, weaving our lives into the larger pattern of existence, shaping and being shaped by the forces that move through us.

The thread of destiny does not bind us in chains, nor does it lead us blindly through the labyrinth of life. It is more like a guide, a subtle pull that beckons us toward certain experiences, certain encounters, certain moments of transformation. It is the call of the soul, the whisper of the universe that guides us toward the fulfilment of our potential, toward the realization of the deeper truths that lie at the heart of existence. Yet, within this call, there is always the freedom to choose, to shape the path we walk, to decide how we will respond to the moments that unfold before us.

Free will is the power that allows us to shape our lives, to make choices that reflect our deepest desires, our highest aspirations, our most intimate truths. It is the force of creation, the energy that allows us to bring our inner visions into the outer world, to manifest the reality we wish to experience. And yet, this freedom is not absolute, for it exists within the larger context of the universe, within the intricate web of destiny that connects all beings, all moments, all possibilities. Our choices are not made in isolation; they ripple outward through the field of existence, touching everything, influencing everything, shaping the world around us in ways we cannot always see or understand.

This interplay between free will and destiny, between choice and fate, is the heart of the human experience. We are beings who exist within a universe that is both open and closed, both fixed and fluid, both determined and free. The choices we make shape the path we walk, but the path itself is not entirely of our making. There are forces that guide us, forces that shape our lives in ways we cannot control, forces that move through the currents of time and space like invisible winds, carrying us toward the moments we are meant to experience, the lessons we are meant to learn, the transformations we are meant to undergo.

These forces are not external to us, not imposed upon us by some distant, indifferent universe. They are part of us, part of the deeper currents of being that flow through the soul, part of the larger web of existence that connects all things. Destiny is not something that happens to us; it is something we participate in, something we co-create with the universe, with the forces of time, space, and consciousness. The choices we make are the threads we weave into the larger pattern, the actions we take are the ripples we send through the field of existence, shaping the reality we experience, shaping the destiny that unfolds before us.

But what is this destiny that beckons us forward, that calls us to certain paths, certain moments, certain experiences? Is it a fixed and unchangeable thing, or is it something more fluid, something that shifts and changes in response to the choices we make, the thoughts we think, the desires we hold? The truth lies somewhere in between, for destiny is both fixed and fluid, both predetermined and open to change. There are certain moments, certain experiences, certain encounters that are written into the fabric of our lives, moments that we are meant to experience, lessons we are meant to learn. These moments are like signposts along the path, markers that guide us toward the fulfilment of our deeper purpose, our soul's calling.

But within these moments, there is always the freedom to choose, the freedom to shape the path we walk, the freedom to decide how we will respond to the unfolding of our lives. Free will is the power that allows us to create within the context of destiny, to shape our lives in ways that reflect our deepest truths, our highest aspirations, our most intimate desires. It is the force that allows us to bring our inner vision into the outer world, to manifest the reality we wish to experience. Yet, this freedom is always held within the larger context of the universe, within the intricate web of destiny that connects all beings, all moments, all possibilities.

This web is not a rigid structure, not a fixed and unchanging thing, but a living, breathing field of energy and consciousness, a dynamic interplay of forces that shape and are shaped by the choices we make. The thread of destiny is not a single line, not a path that leads us unerringly toward a predetermined end. It is a complex web of possibilities, a field of potential that shifts and changes in response to the energy we bring to it, the choices we make, the desires we hold. We are not passive travellers on the road of life; we are active participants in the creation of our destiny, co-creators with the universe, shaping the path we walk with every thought, every action, every breath.

This understanding brings with it a profound sense of responsibility, for it reveals that we are not merely victims of fate, not merely puppets of destiny, but co-creators of reality. The choices we make, the actions we take, the energy we bring to the world—these are the threads we weave into the larger fabric of existence, the ripples we send through the field of consciousness that shapes the world around us. We are not separate from the universe; we are part of it, connected to it by the invisible thread of destiny that winds its way through the currents of time and space, connecting all things in its silent embrace.

But this understanding also brings with it a sense of humility, for it reveals that we are not the sole creators of our destiny, not the masters of our fate. There are forces at work in the universe that we cannot control, forces that shape our lives in ways we cannot see or understand. These forces are not separate from us, not imposed upon us by some distant, indifferent universe, but are part of the deeper currents of being that flow through the soul, part of the larger web of existence that connects all things. To live with this awareness is to recognize that we are both free and bound, both creators and created, both shapers of our destiny and participants in the larger flow of existence.

In the ancient traditions, this interplay between free will and destiny was understood and revered. The weavers of fate, the spinners of the cosmic thread, the gods who guided the paths of mortals—they were not seen as tyrants or despots, but as partners in the dance of creation, as forces that worked in harmony with the choices of beings, shaping and being shaped by the energy of the universe. To live with this understanding is to see that destiny is not something to be feared, not something to be resisted, but something to be embraced, something to be danced with, something to be co-created with the universe.

In this dance, free will is not the rejection of destiny, but its fulfilment. It is the power that allows us to shape our lives within the context of the larger pattern, to bring our unique vision, our unique energy, our unique self to the unfolding of the cosmos. Free will is the power that allows us to create within the field of destiny, to shape the path we walk, to decide how we will respond to the moments that unfold before us. And destiny, in turn, is not the negation of free will, but its partner, its guide, its silent companion that beckons us toward the fulfilment of our deeper purpose, our soul's calling.

In the quiet moments of reflection, we may catch a glimpse of this deeper reality, a flicker of the thread of destiny that winds its way through the currents of time and space, connecting all things in its silent embrace. We may come to understand that fate and free will are not opposites, but partners in the dance of creation, that destiny is not something that happens to us, but something we participate in, something we co-create with the universe. And in this understanding, we may find a new way of being, a new way of living, a new way of understanding ourselves and the world around us— a way that embraces the delicate thread of destiny, the infinite potential of free will, and the interwoven path of being that winds its way through the cosmos, guiding us toward the fulfilment of our deepest purpose, our highest truth, our most intimate self.

CHAPTER 26: THE HIDDEN GEOMETRY OF THE SOUL

Symmetry, Balance, And The Sacred Design Of Being

There exists a hidden geometry within the soul, an unseen architecture that shapes the very essence of our existence. It is a geometry that moves not with lines and angles but with the fluid grace of the infinite—a symmetry of being that transcends the physical world, weaving together the visible and the invisible, the finite and the eternal. This geometry is not etched upon the surface of our lives but flows through the depths of our consciousness, guiding the patterns of thought, emotion, and experience that shape our reality. To awaken to this geometry is to glimpse the sacred design of the universe, to perceive the delicate balance that holds all things in harmony, to feel the pulse of creation moving through the core of our being.

In the world around us, symmetry and balance are everywhere, though often unnoticed. The spirals of galaxies mirror the curls of seashells; the branching of trees echoes the fractal patterns of rivers and lightning. Even within the cells of our bodies, the same sacred proportions guide the replication of life, the unfolding of complexity from simplicity. This is not mere coincidence but the reflection of a deeper order, a cosmic harmony that moves through all things, connecting the smallest particle to the vastest star. It is an order that speaks of unity, of a grand design in which all things are interconnected, woven together by the same invisible threads of energy and meaning.

Yet, for all its elegance, this geometry is not rigid or fixed. It is alive, breathing, moving with the rhythm of creation and dissolution, expansion and contraction. The symmetry of the universe is not the symmetry of the static, of the unchanging. It is a dynamic balance, a harmony that flows and shifts with the tides of existence, constantly adapting, constantly evolving. Just as a dancer moves with grace and fluidity, balancing on the edge of motion, so too does the universe maintain its symmetry through the endless dance of forces—light and darkness, creation and destruction, being and non-being. This dance is the heart of the sacred geometry, the pulse of the cosmos that moves through all things, giving form to the formless, order to the chaos, balance to the infinite.

But what of the soul? What of the inner world of consciousness that moves within each of us, shaping our thoughts, our emotions, our experiences? Does this hidden geometry extend into the depths of our being, guiding the unfolding of our inner lives as it does the patterns of the physical world? The answer, perhaps, lies in the way we experience balance and harmony in our own lives—the moments when everything feels aligned, when our thoughts, emotions, and actions move in perfect synchrony, when we feel a deep sense of connection to the world around us. These moments are rare, fleeting, like the brief alignment of the planets in the night sky, but they offer a glimpse into the deeper symmetry that moves through the soul, the hidden geometry that shapes the unfolding of our consciousness.

This symmetry is not something that can be grasped by the intellect alone, for it is not a matter of logic or reason. It is a matter of being, a matter of feeling the flow of energy within ourselves and aligning with the greater flow of the universe. The hidden geometry of the soul is the balance between the inner and the outer, between the self and the world, between thought and action, between desire and fulfilment. It is the harmony that arises when we move with the rhythm of life, when we allow ourselves to be guided by the deeper currents of consciousness, when we surrender to the flow of existence and trust in the balance that lies at the heart of all things.

But this balance is not static; it is not something that can be achieved and then maintained indefinitely. It is a dynamic process, a constant adjustment, a continual dance between opposites, between forces that pull us in different directions. The soul, like the universe, is always in motion, always evolving, always seeking to maintain its balance even as it moves through the cycles of growth, change, and transformation. To live in harmony with this hidden geometry is to embrace the flow of life, to recognize that balance is not something to be achieved but something to be experienced, something to be lived.

This hidden geometry reveals itself not only in the moments of harmony and balance but also in the moments of chaos and disarray. For just as there is beauty in symmetry, there is beauty in asymmetry, in the unpredictable, in the unexpected twists and turns of life. The sacred geometry of the soul is not a rigid structure, not a perfect alignment of forces, but a fluid and dynamic process, a dance between order and chaos, between symmetry and asymmetry. The soul, like the universe, moves through cycles of balance and imbalance, through moments of clarity and confusion, through periods of peace and turmoil. Yet, beneath it all, there is a deeper order, a hidden symmetry that holds everything together, that guides the unfolding of our lives even when we cannot see it.

This understanding brings with it a sense of peace, for it reveals that the moments of imbalance, the times when we feel lost or uncertain, are not failures but part of the larger process of growth and transformation. The hidden geometry of the soul is not about maintaining perfect balance at all times but about moving with the rhythm of life, about trusting that the moments of chaos and confusion are just as important as the moments of clarity and alignment. To live with this awareness is to embrace the full spectrum of existence, to recognize that balance is not a destination but a journey, a continual process of adjustment and realignment.

In this vision of the soul, there is no separation between the inner and the outer, between the self and the universe. The hidden geometry that moves through the cosmos moves through us as well, shaping the patterns of our thoughts, our emotions, our actions. We are not separate from the world around us but are part of the same sacred design, part of the same cosmic harmony. The balance we seek in our lives is not something that can be achieved through effort or willpower but something that arises naturally when we align ourselves with the deeper flow of existence, when we surrender to the rhythm of life and trust in the balance that lies at the heart of all things.

This understanding brings with it a new way of seeing the self, a new way of understanding identity. The self is not a fixed or static thing, not a singular entity that exists apart from the world. The self is a dynamic process, a wave of consciousness that moves through the hidden geometry of the soul, shaped by the same forces that shape the universe. To be fully alive is to embrace this fluidity, to recognize that we are not defined by our thoughts, our emotions, or our actions, but by the deeper currents of consciousness that move through us, by the hidden symmetry that guides the unfolding of our being.

In the ancient traditions, this hidden geometry was understood and revered. The mandalas of the East, the sacred circles of the West, the labyrinths of old—they were all expressions of the same deeper truth, the same recognition of the sacred design that moves through all things. These symbols were not merely decorative or artistic but were reflections of the hidden geometry that shapes the universe, the sacred patterns that guide the flow of energy and consciousness. To contemplate these symbols was to connect with the deeper order of existence, to feel the pulse of the cosmos moving through the self, to align with the sacred design of being.

But this hidden geometry is not confined to the symbols of the ancient traditions; it is present in every aspect of life, in every moment of awareness, in every breath we take. The sacred design that moves through the stars moves through us as well, shaping the patterns of our lives, guiding the flow of energy and consciousness that moves through the soul. To live with this awareness is to recognize that we are part of something much larger than ourselves, part of the same cosmic dance that moves through the universe. The balance we seek is not something that can be achieved through effort or control but something that arises naturally when we align ourselves with the deeper flow of existence, when we surrender to the rhythm of life and trust in the hidden geometry that guides the unfolding of our being.

In the quiet moments of reflection, we may catch a glimpse of this deeper reality, a flicker of the hidden geometry that moves through the soul, a glimmer of the sacred design that shapes the universe. We may come to understand that balance is not something to be achieved but something to be experienced, something to be lived. And in this understanding, we may find a new way of being, a new way of living, a new way of understanding ourselves and the world around us—a way that embraces the hidden geometry of the soul, the sacred design of being, and the dynamic balance that moves through all things, guiding the flow of existence with grace, harmony, and infinite wisdom.

This balance is not a final state, not an endpoint to be reached, but an ever-present process, a continual unfolding of energy and consciousness that shapes the patterns of our lives. To live with this awareness is to embrace the fluidity of existence, to recognize that we are part of the same sacred geometry that moves through the stars, the trees, the rivers, and the mountains. It is to understand that the balance we seek is not something outside ourselves but something that moves within us, something that guides the unfolding of our being with the same grace and harmony that shapes the universe itself.

The sacred design of being is not a mystery to be solved, but a truth to be lived. It is the hidden geometry that moves through the soul, the dynamic balance that guides the flow of energy and consciousness through all things. To awaken to this geometry is to align with the deeper currents of existence, to trust in the balance that lies at the heart of all things, to feel the pulse of the cosmos moving through the core of our being. And in this alignment, we find not only balance but a deeper sense of connection, a deeper sense of purpose, a deeper sense of being part of something infinitely larger than ourselves, something that transcends the boundaries of time and space, something that moves through the very heart of existence itself.

CHAPTER 27: THE ETHEREAL WEB

The Interconnectedness Of All Things, Time, And The Cosmic Mind

In the quiet spaces between thought and breath, there is a web—an ethereal lattice that stretches beyond the limits of perception, binding every particle, every pulse of energy, and every spark of consciousness. It is not a web made of threads we can see or touch, but a network of connection that exists in the realm of the unseen, where the boundaries between beings dissolve, and the distance between stars becomes as close as a whisper. This web is the foundation of the cosmos, a structure that holds all of creation in delicate balance, a reflection of the infinite unity that weaves through the multiplicity of existence. To perceive this web is to awaken to the truth that nothing exists in isolation, that all things—beings, moments, thoughts—are part of an indivisible whole, moving in harmony with the great cosmic mind that breathes life into the universe.

The ancient mystics, gazing into the night sky, sensed this invisible connection, though they may have called it by different names: the Great Chain of Being, the Eternal Circle, the Divine Mind. It was not something they could measure or describe in material terms, but something they felt, something they knew intuitively, a truth that resonated deep within their souls. The stars above were not distant points of light, but nodes in a vast network of meaning, connected to each other and to the earth, to the waters, to the winds, and to the hearts of those who looked upon them. This understanding of interconnectedness was not confined to the stars; it extended to all things—to the smallest blade of grass, to the stones that lined the rivers, to the thoughts and dreams that drifted through the human mind.

To live is to be part of this web, to feel its threads stretching through time and space, linking every moment, every being, every action in a network of cause and consequence, of presence and potential. It is a web that hums with the energy of creation, that vibrates with the music of existence, that pulses with the rhythm of the universe itself. Yet, for most of us, this web remains hidden, veiled by the illusions of separation, by the belief that we are solitary beings, moving through an external world and apart from us. We see ourselves as individuals, distinct from one another, each with our separate lives, separate desires, and separate destinies. We move through the world as though we were islands, isolated and self-contained, unaware that beneath the surface of our perception, we are all connected, all part of the same vast, living organism.

But there are moments, rare and fleeting, when the veil lifts when the illusion of separation dissolves, and we catch a glimpse of the deeper reality that lies beneath. These moments come to us in different ways—through a sudden insight, a moment of stillness, a profound encounter with nature, or a deep connection with another being. In these moments, we feel the presence of the web, the invisible threads that bind us to all things, the subtle energy that moves through the universe and us. We understand, if only for an instant, that we are not separate, that we are part of something infinitely larger than ourselves, something that transcends the boundaries of time and space, something that connects us to the stars, to the earth, and each other in ways we can scarcely comprehend.

This web of interconnectedness is not merely a philosophical idea, not merely a metaphor for the unity of existence. It is a reality, a fundamental truth of the universe that has been revealed not only through mystical experience but through the discoveries of science. In the quantum world, particles are not isolated objects but are entangled with one another, linked in ways that defy the limits of time and space. A change in one particle can instantaneously affect another, no matter how far apart they may be. This phenomenon, known as quantum entanglement, reveals that at the most fundamental level, the universe is not made up of separate parts but is a single, interconnected whole, where everything is linked, everything is part of the same field of energy and information.

But this interconnectedness extends beyond the quantum realm; it is present in every aspect of existence, from the smallest particle to the largest galaxy, from the flow of time to the fabric of space. The universe is not a collection of isolated things but a web of relationships, a network of connections that stretches through all dimensions of reality. Every thought we think, and every action we take sends ripples through this web, influencing not only our own lives but the lives of others, the course of events, and the unfolding of time itself. We are not passive observers of the universe but active participants in its creation, co-creators of reality who shape the world through our consciousness through the energy we bring to each moment.

This understanding brings with it a profound sense of responsibility, for it reveals that our actions, our thoughts, and our very presence in the world have far-reaching consequences, consequences that ripple outward through the web of existence, touching everything and affecting everything. We are not isolated beings moving through a world that is separate from us; we are part of the same living organism, part of the same cosmic mind that guides the unfolding of reality. To live with this awareness is to recognise that we are connected to all things, that the choices we make, the energy we bring to the world, have the power to shape not only our lives but the lives of others, the course of events, the unfolding of the universe itself.

But this interconnectedness is not something that confines us, not something that limits our freedom. On the contrary, it reveals the infinite potential that lies within each of us, the power we have to shape our reality to create the world we wish to experience. The web of existence is not a rigid structure, not a fixed and unchanging thing, but a dynamic, living field of energy and consciousness, a field that responds to the vibrations we send into it, that shifts and adapts to the choices we make, the thoughts we think, the desires we hold. We are not bound by the web; we are part of its creation, part of its unfolding, part of its endless dance of becoming.

This dance of interconnectedness is not something that happens outside of us but something that happens through us. The web of existence is not separate from the self, not something that exists apart from our consciousness. It reflects the cosmic mind, a reflection of the infinite intelligence that moves through the universe and us. To awaken to this truth is to awaken to the deeper reality that lies beneath the surface of our perception, to feel the pulse of the universe moving through our being, to recognise that we are not separate from the world but are part of the same infinite consciousness, part of the same eternal mind that shapes the stars, the galaxies, the atoms, and the thoughts that drift through our awareness.

In this vision of reality, time and space are not fixed dimensions, not rigid structures that confine us to a particular place or moment. They are fluid, malleable, part of the same dynamic web of existence that connects all things. Time is not a straight line, moving from past to future in a relentless march toward some distant goal. It is a field, a dimension of possibility that shifts and bends in response to consciousness, to the choices we make, and to the energy we bring to each moment. The future is not something that lies ahead of us, waiting to be reached; it is something that is constantly being created, constantly emerging from the infinite potential that lies within the present moment. The past is not something that is fixed and unchangeable; it is part of the same dynamic field of time, part of the same web of existence that connects all moments, all beings, all things.

This understanding brings with it a sense of freedom, for it reveals that we are not bound by the past, not confined by the limitations of time and space. We are free to create, free to shape our reality, free to move through the web of existence with the knowledge that we are part of something much larger than ourselves, part of the same cosmic dance that moves through the stars, the galaxies, the atoms, and the thoughts that drift through our consciousness. To live with this awareness is to recognise the infinite potential that lies within each moment, to understand that the choices we make, the energy we bring to the world, have the power to shape not only our own lives but the course of the universe itself.

This interconnectedness is not something we can fully understand with the mind alone, for it is not a concept or a theory but an experience, a truth that can only be felt, known in the deepest parts of our being. To awaken to this truth is to feel the pulse of the universe moving through us, to feel the threads of connection that bind us to all things, to experience the unity of existence in a way that transcends thought, transcends language, transcends the boundaries of the self. It is to realise that we are not separate from the world but are part of the same infinite consciousness, part of the same cosmic mind that shapes the unfolding of reality.

In the quiet moments of reflection, we may catch a glimpse of this deeper reality, a flicker of the web of existence that stretches through time and space, connecting all things in its silent embrace. We may come to understand that we are not isolated beings, moving through a world that is separate from us but are part of the same living organism, part of the same cosmic mind that guides the unfolding of the universe. And in this understanding, we may find a new way of being, a new way of living, a new way of understanding ourselves and the world around us—a way that embraces the ethereal web of interconnectedness, the fluid nature of time and space, and the infinite potential of the cosmic mind that moves through all things, guiding the dance of creation with grace, wisdom, and infinite love.

CHAPTER 28: THE SILENT SEED OF TRANSFORMATION

Change, Becoming, And The Eternal Unfolding Of Life

Beneath the surface of existence, beyond the reach of time's relentless passage, there lies a silent seed, an invisible force that stirs in the depths of all things. This seed is the essence of transformation, the catalyst of change, the spark that ignites the eternal process of becoming. It is not bound by the laws of matter or confined to the limits of form; it moves freely through the fabric of reality, shaping and reshaping the world in ways both subtle and profound. This seed of transformation is woven into the very fabric of life itself, a thread that connects every being, every moment, and every experience and guides the endless unfolding of existence with an unseen hand.

In the natural world, we witness the beauty of transformation in the rhythm of the seasons, in the metamorphosis of the caterpillar into the butterfly, and the slow turning of the leaves from green to gold. Yet, these changes are but the visible expression of a much deeper process, a process that moves beneath the surface of things, shaping the very essence of life. The seed of transformation lies not only in the physical realm but in the realm of consciousness, in the inner world where thought, emotion, and spirit merge to create the dance of becoming. This dance is not linear, not a straight path from one state to another, but a spiralling, cyclical process, one that unfolds with infinite complexity and grace, guiding each being toward its highest potential, toward the realisation of its true nature.

To live is to be in a constant state of transformation, to be always becoming, always unfolding, always reaching toward something greater than what we are in this moment. Yet, for most of us, change is something we resist, something we fear. We cling to the familiar, to the known, to the safe, as though the act of becoming were a threat, as though transformation were something to be avoided. But transformation is not a disruption; it is the essence of life itself. It is the force that moves through all things, that breathes new life into the world with every moment, that keeps the universe in a state of perpetual creation. To resist transformation is to resist life, to deny the very nature of existence, to turn away from the process of becoming that is the birthright of every living being.

There is a stillness that precedes transformation, a quiet space where the old begins to dissolve, and the new has not yet taken shape. This stillness is not emptiness; it is the fertile ground in which the seed of transformation takes root, the silence in which the first stirrings of change can be felt. It is a space of potential, a place where anything is possible, where the future is not yet written, and where the past has been released. To dwell in this stillness is to be on the threshold of transformation, to stand at the point where the old self begins to dissolve, and the new self begins to emerge. It is a moment of great power, a moment of infinite possibility, a moment that contains within it the seeds of all future becoming.

Yet, for all its power, transformation is not something that can be forced. It is not something that can be willed into being through sheer determination or effort. It is a process that unfolds in its own time, according to its rhythm, guided by forces that are beyond our control. The seed of transformation must be nurtured, not by force but by patience, by trust, by surrender to the deeper currents of life that move through us. It is a process that requires faith, faith in the unseen forces that guide the unfolding of our lives, and faith in the intelligence of the universe that knows the perfect time for each seed to bloom.

This process of transformation is not confined to the outer world to the changes we see in our bodies, our circumstances, and our surroundings. It is a process that moves through the inner world as well, shaping our thoughts, our emotions, our beliefs, and our very sense of self. The self, like the universe, is not a fixed or static thing, not a solid entity that exists apart from the flow of time. The self is a process, a wave of consciousness that is always in motion, always evolving, always becoming. To know the self is to know this process, to understand that we are not the same person today as we were yesterday, and we will not be the same person tomorrow as we are today. The self is a dynamic and fluid entity, one that is shaped and reshaped by the forces of transformation that move through us.

This understanding brings with it a sense of liberation, for it reveals that we are not bound by the past, not confined by the limitations of who we have been. We are free to become, free to transform, free to evolve into something greater, something more expansive, something more aligned with the deeper truth of our being. The seed of transformation lies within each of us, waiting for the moment when it will be awakened when the conditions are right for it to begin its journey toward the light. This journey is not always easy, for transformation often requires us to let go of the old, to release the familiar, to surrender to the unknown. But it is through this process of letting go that we make space for the new that we create the fertile ground in which the seed of transformation can take root and grow.

This journey of transformation is not a solitary one, for we are not isolated beings moving through a world that is separate from us. We are part of the same field of energy, the same web of existence, the same dance of becoming that moves through the stars, the planets, the trees, and the rivers. The seed of transformation that lies within us is part of the larger process of transformation that moves through the universe, a process that connects all things and binds all beings together in the eternal dance of creation and dissolution, of birth and rebirth. To transform is not merely to change; it is to participate in this larger process, to become part of the cosmic rhythm that moves through all things, and to align ourselves with the deeper flow of life that guides the unfolding of the universe.

But this alignment is not something that can be achieved through effort or willpower alone. It requires a deep surrender, a letting go of the need to control, and a trust in the forces that move through us and the world. It requires us to listen, to tune in to the subtle whispers of the soul, to follow the inner guidance that leads us toward the fulfilment of our potential, toward the realisation of the deeper truth of our being. This guidance is not always loud or clear; it often comes to us in the form of quiet nudges, gentle whispers, and subtle signs that point the way forward. To follow this guidance is to trust in the process of transformation, to trust that the seed of change that lies within us will bloom in its own time, in its way, according to the wisdom of the universe.

This process of transformation is not about becoming something different, something other than what we are. It is about becoming more of what we are, about realising the fullness of our being, about awakening to the deeper truth that lies within us. The seed of transformation is not something foreign or external; it is the essence of our being, the core of who we are. To transform is not to change into something else but to reveal more fully the truth of who we have always been, to unfold the potential that has always been there, waiting to be realised. It is a process of unveiling, of removing the layers of illusion, of letting go of the false identities we have created and stepping into the light of our true nature.

This understanding brings with it a sense of peace, for

it reveals that we do not need to strive or struggle to become something greater. The greatness we seek is already within us, waiting to be revealed, waiting to be awakened through the process of transformation. The seed of transformation lies dormant within us, waiting for the right conditions the right moment, to begin its journey toward the light. This journey is not one of effort but of surrender, of trust, of allowing the deeper forces of life to guide the unfolding of our being.

In ancient traditions, the process of transformation was understood as a sacred journey, a journey that led not only to physical or material change but to spiritual awakening, to the realisation of the deeper truth of existence. The alchemists spoke of the transformation of base metal into gold, but this was not merely a physical process; it was a metaphor for the transformation of the self, for the awakening of the soul to its true nature. The seed of transformation was seen as the essence of life, the spark of divinity that lies within each of us, waiting to be awakened, waiting to be realised through the process of spiritual growth and evolution.

But this journey of transformation is not a linear path, not a straight line from ignorance to enlightenment, from darkness to light. It is a spiral, a cyclical process that moves through phases of growth and contraction, of clarity and confusion, of light and shadow. The seed of transformation does not grow in a straight line; it unfolds in spirals, in waves, in rhythms that are guided by the deeper currents of life. To walk this path is to embrace the full spectrum of existence, to move with the rhythm of life, and to trust in the process of becoming, even when the way forward is unclear.

In the quiet moments of reflection, we may catch a glimpse of this deeper process, a flicker of the seed of transformation that lies within us, waiting to be awakened. We may come to understand that transformation is not something to be feared or resisted but something to be embraced, something to be welcomed as part of the natural unfolding of life. And in this understanding, we may find a new way of being, a new way of living, a new way of understanding ourselves and the world around us—a way that embraces the silent seed of transformation, the eternal process of becoming, and the infinite unfolding of life that moves through all things, guiding the dance of creation with grace, wisdom, and infinite possibility.

CHAPTER 29: THE UNFOLDING HORIZON

Infinity, Consciousness, And The Eternal Journey Of The Soul

There is a horizon that stretches endlessly before the soul, not a line drawn by the limits of space or time, but an infinite expanse where consciousness wanders, forever unfolding into deeper realms of being. It is not a boundary but a beckoning invitation, a silent summons to explore the farthest reaches of existence, to voyage beyond the known, beyond the finite, into the boundless ocean of infinity. This horizon is not distant, not something that lies beyond our grasp, but is woven into the very essence of consciousness itself, a reflection of the infinite potential that resides within every thought, every breath, and every pulse of life. To contemplate this horizon is to awaken to the profound truth that existence, in its essence, is infinite—that the soul, in its journey through time and space, is not bound by the finite but is always reaching, always expanding, always becoming.

Infinity is a concept that eludes the grasp of the mind, for the mind, conditioned by the rhythms of the material world, seeks boundaries, definitions, and limits. We measure the world in days and years, in distances and destinations, as though existence could be contained, as though reality could be captured within the confines of time and space. Yet, beneath the surface of these measurements, beneath the apparent solidity of form, lies the boundless nature of reality, a truth that transcends all limits, all definitions, and all attempts to quantify or contain it. This truth is not something that can be grasped by the intellect, not something that can be reduced to a formula or equation. It is something that must be felt, experienced, and known in the deepest part of the soul.

To awaken to the infinite is to recognise that life is not a journey toward a destination, not a movement from one point to another, but an eternal unfolding, an endless process of becoming. There is no final destination, no point of arrival, no moment when the journey is complete. The horizon always expands, always beckons us forward, drawing us into deeper realms of understanding, into higher states of consciousness, and greater expressions of being. This eternal unfolding is not something separate from the self, not something that happens to us from the outside. It is the very essence of who we are, the core of our being, the pulse of our consciousness. The soul, like the universe, is infinite, and its journey is one of endless exploration, endless creation, and endless becoming.

But what does it mean to journey into the infinite? What does it mean to move beyond the finite, beyond the boundaries of time and space, and into the boundless realms of consciousness? It is not a journey that can be taken with the body, not a voyage that can be measured in miles or years. It is a journey of the soul, a journey that takes place within the depths of consciousness, within the silent spaces of awareness, where time dissolves, and space expands into the infinite. To journey into the infinite is to let go of the need for boundaries, for definitions, for certainties. It is to surrender to the flow of existence, to trust in the process of becoming, and to embrace the mystery of life in all its vastness and complexity.

This journey is not one of escape, not a flight from the material world or the challenges of life. It is a journey that brings us deeper into life, deeper into the heart of existence, deeper into the truth of who we are. The infinite is not something that exists beyond this world, beyond this life but is woven into the very fabric of reality, present in every moment, in every breath, in every experience. To awaken to the infinite is to see the boundless potential that lies within every moment of existence, to recognise that the universe is not a static or fixed thing but a living, breathing process of creation, constantly unfolding, expanding, constantly becoming.

In ancient traditions, the infinite was often described as a vast ocean, a boundless sea that stretched beyond the horizon, a symbol of the endlessness of existence and the limitless potential of the soul. To voyage into this ocean was to embark on a journey of transformation, a journey that led not to a final destination but to an ever-deepening experience of the self, of the universe, of the divine. The ocean was not a place to be conquered or crossed but a realm to be explored, a mystery to be embraced, a reflection of the infinite depths of consciousness that lie within each of us.

Yet, for all its beauty, the infinite can be a source of fear, for it challenges the mind's need for certainty, stability, and boundaries. We seek comfort in the known, in the familiar, in the finite, as though the infinite were something to be feared, something that would overwhelm or dissolve the self. But the truth is that the self is not diminished by the infinite; it is expanded, enriched, and deepened by the recognition of its boundless nature. The self, like the universe, is not a fixed or finite thing but a dynamic and evolving process, a wave of consciousness that moves through the infinite, shaping and being shaped by the forces of creation that flow through it.

To embrace the infinite is to embrace the fullness of existence, to recognise that life is not a linear journey with a beginning and an end but a cyclical, spiralling process, one that moves through phases of creation, dissolution, and rebirth. The horizon that stretches before us is not a line that marks the boundary of our existence but a reflection of the infinite potential that lies within us, the boundless nature of consciousness that is always expanding, always reaching toward new realms of being. The soul, in its journey through time and space, is not seeking a final destination but is engaged in the eternal process of becoming, of unfolding its infinite potential, of exploring the boundless realms of existence.

This understanding brings with it a profound sense of peace, for it reveals that there is no need to rush, no need to strive for some distant goal or final destination. The journey is not about reaching a point of completion but about experiencing the fullness of life, about embracing the infinite potential that lies within every moment of existence. The horizon is not something to be crossed but something to be experienced, something to be lived. To journey into the infinite is to live fully, to engage with the process of becoming in all its complexity, and to embrace the flow of existence with an open heart and an open mind.

Yet, this journey is not always easy, for the infinite challenges us to let go of the need for control, surrender to the flow of life, and trust in the process of becoming. The mind, conditioned by the material world, seeks boundaries, definitions, and certainties. It clings to the finite, to the known, to the familiar, as though the infinite were a threat, as though the vastness of existence were something to be feared. But the soul, in its essence, knows the truth. It knows that the infinite is not a threat but a source of liberation, a source of expansion, a source of infinite potential. To journey into the infinite is to reconnect with the deeper truth of who we are, to remember that we are not bound by the limitations of time and space but are part of the infinite flow of existence, part of the boundless ocean of consciousness that moves through the universe.

In the quiet moments of reflection, we may catch a glimpse of this deeper reality, a flicker of the infinite that lies within us, waiting to be explored. We may come to understand that life is not a linear path, not a journey toward a final destination, but an endless process of becoming, an eternal unfolding of consciousness into deeper realms of being. The horizon that stretches before us is not a line that marks the boundary of our existence but a reflection of the boundless potential that lies within us, the infinite nature of consciousness that is always expanding, always reaching toward new realms of understanding, new expressions of being.

This understanding brings with it a sense of freedom, for it reveals that we are not bound by the limitations of the material world, not confined by the boundaries of time and space. We are free to explore, free to create, free to become, free to expand into the infinite potential that lies within us. The journey of the soul is not a quest for completion but an exploration of the infinite, an unfolding of consciousness that moves through the boundless realms of existence, shaping and being shaped by the forces of creation that flow through it. To live with this awareness is to embrace the fullness of life, to engage with the process of becoming in all its richness and complexity, and to trust in the infinite potential that lies within every moment of existence.

The infinite is not something that exists beyond this world, beyond this life but is woven into the very fabric of reality, present in every breath, in every thought, in every experience. To awaken to the infinite is to see the boundless nature of existence, to recognise that life is not a finite journey but an eternal unfolding, an endless process of becoming. The horizon that stretches before us is not a destination to be reached but a reflection of the infinite potential that resides within us, the boundless nature of consciousness that is always expanding, always reaching toward new realms of being.

In the quiet moments of stillness, we may feel the pull of this horizon, the silent summons to explore the infinite depths of consciousness, to venture beyond the known, beyond the finite, into the boundless realms of existence. We may come to understand that life is not a journey toward a final destination but an endless process of becoming, an eternal unfolding of consciousness into deeper realms of being. And in this understanding, we may find a new way of being, a new way of living, a new way of understanding ourselves and the world around us —a way that embraces the infinite nature of existence, the boundless potential of consciousness, and the eternal journey of the soul as it moves through the endless horizon of life, forever becoming, unfolding, forever reaching toward the infinite.

CHAPTER 30: THE BREATH OF THE ETERNAL

Unity, Wholeness, And The Mystery Of Being

There is a breath that moves through all things, a silent exhalation that echoes through the corridors of time and space, touching every atom, every soul, every flicker of consciousness. It is not the breath of the body, not the rhythmic rise and fall of lungs drawing air, but the breath of existence itself—the pulse of the eternal that weaves through the fabric of reality, binding all things in an indivisible whole. This breath is the essence of unity, the invisible thread that connects the stars to the earth, the seen to the unseen, the known to the unknown. To breathe is to participate in this cosmic rhythm, to feel the pulse of creation moving through the self, to awaken to the truth that all things, all beings, and all moments are part of the same infinite, unbroken whole.

To contemplate unity is to step beyond the boundaries of the self, to dissolve the illusion of separation that defines our ordinary perception of the world. We live in a world of divisions—between the self and the other, between mind and body, between spirit and matter. These divisions shape the way we experience reality, creating the illusion that we are separate from the world, from each other, from the universe itself. But beneath these divisions lies a deeper truth, a truth that whispers through the silence of being, reminding us that we are not separate, that we are part of the same vast, interconnected field of existence, and that the boundaries we perceive are illusions born of limited perception.

The mystics of old spoke of this unity in reverent tones, describing it as the oneness of all things, the divine harmony that moves through the cosmos, the sacred wholeness that holds the universe in balance. It was not something they could see with their eyes but something they felt in the depths of their being, something they knew intuitively, a truth that transcended the limits of language and thought. For them, unity was not an abstract concept, not a philosophical idea, but a living, breathing reality, a reality that could be experienced directly through the expansion of consciousness, through the dissolution of the ego, through the merging of the self with the infinite.

This unity is not something that exists outside of us, not something that we must seek or strive to attain. It is the essence of who we are, the core of our being, the foundation of all existence. To awaken to this unity is not to discover something new but to remember something ancient, something primordial, something that has always been present, hidden beneath the surface of our awareness. It is to remember that we are not separate from the universe but are expressions of it, that the same breath that moves through the stars moves through us, and that the same force that shapes the galaxies shapes our thoughts, our emotions, and our lives.

Yet, for most of us, this unity remains hidden, veiled by the illusions of separation that dominate our perception of the world. We see ourselves as individuals, as discrete entities, moving through a world that is external and apart from us. We define ourselves by our differences, by the boundaries that separate us from each other, from the world, from the divine. We cling to these boundaries, to the sense of self that gives us a feeling of solidity, of identity, of control. But the self, like all things, is not a fixed or separate entity; it is a process, a wave of consciousness that moves through the field of existence, shaped and reshaped by the forces of creation that flow through it.

To awaken to the unity of all things is to dissolve the boundaries of the self, to step beyond the illusion of separation, to merge with the infinite flow of existence that moves through the universe and us. It is to recognise that the self is not an island, not a solitary being, but a wave in the ocean of consciousness, a moment of awareness that arises from the infinite and dissolves back into it, only to arise again in a new form, a new expression of the eternal dance of being. The self is not separate from the world but is part of the same cosmic rhythm, part of the same breath of existence that flows through all things.

This understanding brings with it a sense of profound peace, for it reveals that we are not alone, not isolated beings struggling against the forces of life, but are part of the same infinite, unbroken whole. The challenges we face, the struggles we endure, the joys we experience—these are not isolated events, not random occurrences, but are part of the larger flow of existence, part of the same cosmic dance that moves through the stars, through the earth, through the hearts and minds of all beings. To live with this awareness is to recognise that we are not separate from the world but are part of it, connected to it in ways that are both intimate and profound.

But this unity is not something that negates the individuality of the self, not something that dissolves the unique expression of consciousness that each of us represents. On the contrary, it reveals that individuality and unity are not opposites, not contradictory forces, but are part of the same whole, part of the same infinite field of being. The self, in its unique expression, reflects the infinite, a moment of awareness that arises from the same source, the same breath, the same cosmic mind that shapes the universe. To embrace unity is not to lose the self but to recognise that the self is part of the whole, that the individual is an expression of the universal, and that we are both distinct and one, both unique and inseparable from the greater field of existence.

This understanding brings with it a new way of seeing the world, a new way of experiencing life. The boundaries that once defined our experience of reality dissolve, revealing the deeper connections that lie beneath the surface of things. We begin to see the world not as a collection of separate objects but as a living, breathing whole, a web of relationships in which all things are connected, all things are part of the same cosmic dance. The trees, the rivers, the mountains, the stars—they are not separate from us, not distant or apart, but are part of the same field of existence, part of the same breath of life that moves through all things.

This unity extends beyond the physical world, beyond the realm of matter and form. It reaches into the realms of thought, emotion, and spirit, connecting us in ways that transcend the limitations of the material world. The thoughts we think, the emotions we feel, the energy we bring to the world—they are not confined to the self, not limited by the boundaries of the body or the mind, but ripple outward through the field of existence, touching everything, affecting everything. We are not separate from each other, not isolated beings moving through a world of separation, but are part of the same field of consciousness, part of the same web of life that connects all beings, all moments, and all things.

This understanding of unity brings with it a sense of responsibility, for it reveals that the choices we make, the actions we take, the energy we bring to the world—these are not isolated events but are part of the larger flow of existence, part of the same cosmic dance that shapes the universe. We are not passive observers of life but active participants in the creation of reality, co-creators of the world we experience. The unity of all things means that we are not separate from the world, not separate from each other, not separate from the divine. To live with this awareness is to recognise the power we hold, the power to shape our reality, to create the world we wish to experience and to bring forth the harmony, the beauty, the love that lies at the heart of existence.

But this power is not something that can be wielded through force or control. It is a power that arises from alignment with the deeper currents of life, from the recognition of our place within the whole, from the understanding that we are part of the same breath of existence that moves through the stars, through the earth, and the hearts of all beings. To live in alignment with this unity is to move with the rhythm of life, to trust in the flow of existence, to embrace the wholeness of being in all its complexity, in all its mystery, in all its beauty.

In the quiet moments of reflection, we may catch a glimpse of this deeper reality, a flicker of the unity that lies beneath the surface of existence, a glimmer of the infinite wholeness that connects all things. We may come to understand that we are not separate from the world but are part of it, that the boundaries we perceive are illusions, that the self is not an isolated being but a wave in the ocean of consciousness, a moment of awareness that arises from the infinite and dissolves back into it. And in this understanding, we may find a new way of being, a new way of living, a new way of understanding ourselves and the world around us—a way that embraces the unity of all things, the wholeness of existence, and the mystery of being that lies at the heart of life.

This mystery is not something to be solved, not something to be understood or explained. It is something to be experienced, something to be lived. The breath of the eternal is not a concept, not a philosophy, but a reality, a living truth that moves through the heart of existence, connecting all things in a web of infinite relationships, in a dance of becoming that has no beginning and no end. To live with this awareness is to breathe with the universe, to move with the rhythm of life, to embrace the wholeness of being in all its forms, in all its expressions, in all its infinite possibilities.

And so, as we come to the close of this chapter, we do not come to an end but to a new beginning, a new moment in the eternal unfolding of life. The horizon stretches ever onward, the breath of the eternal moves through us, and the journey of the soul continues, forever expanding, becoming, forever reaching toward the infinite.

EPILOGUE

The Silent Return To The Source

There is a moment, not measured by the ticking hands of clocks or the turning of the seasons, when the soul, having wandered through the labyrinth of life, pauses. It is a stillness that holds no weight of time, a breathless space where the boundaries of all that was, all that is, and all that will be dissolved into a singular essence. This moment is not the conclusion of a journey, nor is it a destination reached after miles of searching. It is a return—not to a place, but to a state of being, a return to the source from which all things arise. It is the point where the circle of existence folds back upon itself, revealing that the path we walk has always led us home.

What is this source, this origin to which all beings must eventually return? It is not a physical place, not a realm that lies beyond the stars or at the end of time. It is the formless, timeless core of all things, the infinite ground of being from which the universe emerges and into which it dissolves. The source is the silent pulse that moves through the heart of existence, the breath that animates the cosmos, the unbroken stillness from which all movement arises. It is the eternal presence that lies beneath the play of form and formlessness, the unity that holds the many in its embrace, the one that sings through the countless expressions of life.

To speak of the source is to speak of the mystery that lies at the heart of all things—a mystery that cannot be named, cannot be grasped, and cannot be held within the confines of language or thought. It is the great paradox of existence, the unmanifest that gives rise to the manifest, the unseen that gives birth to the seen, the silence that underlies all sound. In the course of our lives, we are drawn outward into the world of form, into the dance of creation, into the play of light and shadow, of birth and death, of joy and sorrow. We become mesmerised by the movement, by the constant unfolding of life, and in this dance, we forget the stillness that holds it all, the source from which it arises.

Yet, no matter how far we wander, no matter how deeply we become entangled in the drama of existence, there is always a part of us that remembers. There is always a part of the soul that is inextricably linked to the source, a part that never left, never forgot, never ceased to feel the silent pulse of the eternal moving through it. This part of us, often hidden beneath the layers of thought, emotion, and experience, quietly guides us back, calling us to remember the truth of who we are to return to the essence of being that lies beyond the changing forms of life.

The journey of the soul, then, is not a quest for something

new, not a search for some distant truth. It is a return—a return to the source, a return to the unchanging essence that lies beneath the shifting patterns of existence. It is a return to the stillness that holds all movement, to the silence that underlies all sound, and to the unity that embraces all duality. This return is not a retreat from life, not an escape from the world, but a deepening into the heart of existence, a recognition that the source is not something separate from the world but is woven into the very fabric of reality. To return to the source is to recognise that we are not separate from life, not separate from the universe, not separate from each other, but are part of the same infinite whole, the same eternal being that moves through all things.

This return is not an event, not a single moment of awakening or realisation. It is a process, a gradual unfolding, a continuous deepening into the truth of who we are. It is the journey of a lifetime, and yet it is a journey that exists outside of time, for the source is not something that can be reached through the passage of time or the accumulation of experience. It is always present, always here, always now, waiting for us to turn our attention inward, to quiet the mind, to still the restless seeking, and to recognise the truth that has always been.

To live with this awareness is to live in harmony with the rhythm of the universe, to move with the flow of existence rather than against it. It is to recognise that life is not something that happens to us, not something that we must struggle to control or understand. Life is a process of unfolding, a continuous expression of the source, a dance of being and becoming that is guided by the same intelligence that moves the stars, shapes the galaxies, and breathes life into every corner of the cosmos. To live with this awareness is to trust in the process of life, surrender to the flow of existence, and embrace the mystery that lies at the heart of all things.

But this surrender is not a passive resignation; it is an active participation in the dance of creation, a conscious alignment with the deeper currents of life that move through us. To return to the source is not to withdraw from the world but to engage with it more fully, to recognise that every moment, every experience, every thought and emotion is an expression of the same infinite being, the same eternal presence that flows through all things. It is to see the sacred in the ordinary, the divine in the mundane, the infinite in the finite.

This recognition brings with it a profound sense of peace, for it reveals that there is no need to strive, no need to search for meaning or purpose outside of ourselves. The meaning of life is not something to be found in the future, not something that lies beyond our current experience. It is here, now, in the silent presence of the source that moves through every moment, through every breath, through every heartbeat. To live in this awareness is to rest in the stillness of being, to trust in the unfolding of life, and to recognise that we are always held by the source, always guided by the same intelligence that shapes the universe.

Yet, even as we return to the source, we continue to participate in the unfolding of creation, for the source is not static, not unchanging, but is a living, breathing presence that moves through the flow of existence. The return to the source is not the end of the journey but the beginning of a new way of being, a way of living in harmony with the rhythm of life, a way of moving through the world with grace, wisdom, and love. It is a way of being that is grounded in the recognition of our unity with all things, our connection to the infinite, and our place within the great web of existence.

In the quiet moments of reflection, we may feel the pull of the source, the silent call to return to the essence of being, to remember the truth of who we are. We may come to understand that life is not a journey toward a final destination but a continuous unfolding, a return to the source that lies at the heart of all things. And in this understanding, we may find a new way of being, a new way of living, a new way of experiencing the world—a way that embraces the unity of existence, the wholeness of life, and the mystery of being that lies at the heart of the cosmos.

As the breath of the eternal moves through us, we are reminded that we are not separate from the source, not separate from each other, not separate from the universe. We are part of the same infinite whole, the same eternal presence that moves through all things. To return to the source is to return to the truth of who we are, to remember that we are not isolated beings moving through a world of separation but are expressions of the same infinite consciousness, the same divine essence, the same breath of existence that flows through the stars, through the earth, through the hearts and minds of all beings.

And so, as we return to the source, we do not leave the world behind but embrace it more fully, recognising that the source is not something that lies beyond the world, beyond life, but is the essence of life itself. The journey of the soul is not a journey away from the world but a journey into the heart of existence, a return to the source from which all things arise and into which all things dissolve. This return is not the conclusion of a story but the opening of a new chapter, a new moment in the eternal unfolding of life. The breath of the eternal moves through us, and the journey continues—forever expanding, forever becoming, forever returning to the source.

REFERENCES

Though the book "The Quantum Soul: Exploring the Metaphysics of Consciousness" is an original work, it draws loose inspiration from a rich tapestry of philosophical, scientific, and spiritual literature. Below is a list of influential works, both contemporary and classical, that have informed the themes, ideas, and conceptual frameworks explored in the book:

1. Capra, Fritjof. The Tao of Physics: An Exploration of the Parallels Between Modern Physics and Eastern Mysticism. Shambhala Publications, 1975.
   - This book explores the connections between quantum physics and Eastern spiritual traditions, a major theme woven throughout the book.

2. Bohm, David. Wholeness and the Implicate Order. Routledge, 1980.
   - Bohm's vision of the universe as an interconnected whole and his concept of the "implicate order" directly influenced the ideas of cosmic unity and interconnectedness presented in the book.

3. Penrose, Roger. The Emperor's New Mind: Concerning Computers, Minds, and the Laws of Physics. Oxford University Press, 1989.
   - Penrose's exploration of consciousness as related to the fundamental laws of physics offers insight into the

quantum nature of the mind, a key theme in this work.

4. Tolle, Eckhart. The Power of Now: A Guide to Spiritual Enlightenment. New World Library, 1997.

- Tolle's reflections on the nature of consciousness, time, and the eternal present inspired the chapters focused on time and the present moment as dimensions of spiritual awakening.

5. Wilber, Ken. The Spectrum of Consciousness. Theosophical Publishing House, 1977.

- Wilber's integral theory and exploration of the levels of human consciousness have deeply influenced discussions of self-awareness, transformation, and the expansion of consciousness.

6. Jung, Carl G. The Archetypes and the Collective Unconscious. Princeton University Press, 1959.

- Jung's exploration of the collective unconscious and his archetypal theory provide the groundwork for understanding the deeper dimensions of the psyche, which are explored in relation to metaphysics and the cosmos.

7. Teilhard de Chardin, Pierre. The Phenomenon of Man. Harper & Row, 1955.

- Teilhard de Chardin's vision of the evolutionary process as a movement toward higher consciousness and the concept of the "Omega Point" resonate with the book's exploration of consciousness as an evolving phenomenon.

8. Kastrup, Bernardo. The Idea of the World: A Multi-Disciplinary Argument for the Mental Nature of Reality. John Hunt Publishing, 2019.

- Kastrup's idealist perspective, which posits that reality is fundamentally mental in nature, underpins the metaphysical framework of consciousness explored in the book.

9. Nagarjuna. The Fundamental Wisdom of the Middle Way (Mulamadhyamakakarika). Translated by Jay L. Garfield, Oxford University Press, 1995.
   - Nagarjuna's philosophical inquiries into emptiness and the nature of reality echo throughout the book's discussions of the formless and the relationship between existence and consciousness.

10. Heisenberg, Werner. Physics and Philosophy: The Revolution in Modern Science. Harper & Brothers, 1958.
    - Heisenberg's thoughts on quantum mechanics and its philosophical implications have been critical in shaping the book's treatment of uncertainty and the observer's role in shaping reality.

11. Hawking, Stephen. A Brief History of Time: From the Big Bang to Black Holes. Bantam Books, 1988.
    - The interplay between time, space, and consciousness in this work is inspired by Hawking's reflections on the nature of the universe and the boundaries of time.

12. Varela, Francisco, Evan Thompson, and Eleanor Rosch. The Embodied Mind: Cognitive Science and Human Experience. MIT Press, 1991.
    - The embodied cognition approach, blending cognitive science and Buddhist philosophy, offers a foundation for the book's inquiry into how consciousness interacts with the physical world.

13. Huxley, Aldous. The Perennial Philosophy. Harper & Brothers, 1945.

- Huxley's exploration of the common threads in mystical and metaphysical traditions throughout history provides a background for the book's universalist approach to consciousness.

14. Laszlo, Ervin. Science and the Akashic Field: An Integral Theory of Everything. Inner Traditions, 2004.

- Laszlo's idea of the Akashic field as a cosmic memory field is foundational to the discussions of memory, time, and the interconnectivity of consciousness explored in the book.

15. Barad, Karen. Meeting the Universe Halfway: Quantum Physics and the Entanglement of Matter and Meaning. Duke University Press, 2007.

- Barad's theory of agential realism, which integrates quantum physics and philosophical inquiry, inspires the book's approach to the relationship between consciousness, matter, and the observer.

16. Rupert Sheldrake. The Science Delusion: Freeing the Spirit of Enquiry. Coronet, 2012.

- Sheldrake's notion of morphic resonance, suggesting that memory is inherent in nature, influenced the exploration of how consciousness and the universe may share fields of collective memory.

17. Krishnamurti, Jiddu. The First and Last Freedom. Harper & Brothers, 1954.

- Krishnamurti's discussions on the nature of freedom, truth, and perception offer a spiritual and philosophical basis for the book's inquiry into the liberating nature of expanded consciousness.

18. Spinoza, Baruch. Ethics. Translated by Edwin Curley, Penguin Classics, 1996.

 - Spinoza's monistic philosophy, which posits that there is only one substance (God or Nature), inspired reflections on the unity and interconnectedness of all things.

19. Einstein, Albert. Relativity: The Special and General Theory. Henry Holt and Company, 1916.

 - Einstein's work on relativity has had a profound influence on the book's treatment of time, space, and the relationship between the observer and the universe.

20. Plotinus. The Enneads. Translated by Stephen MacKenna, Penguin Classics, 1991.

 - Plotinus' vision of the One and the emanation of all things from this singular source provides the philosophical groundwork for the book's metaphysical inquiry into the origin and nature of consciousness.

21. Schrödinger, Erwin. What is Life? With Mind and Matter and Autobiographical Sketches. Cambridge University Press, 1967.

 - Schrödinger's reflections on life, consciousness, and the nature of reality have greatly informed the metaphysical ideas regarding the quantum nature of the soul.

22. Patanjali. The Yoga Sutras of Patanjali. Translated by Swami Satchidananda, Integral Yoga Publications, 1978.

 - The exploration of consciousness and the mind's nature draws inspiration from Patanjali's classical teachings on the layers of consciousness and the ultimate goal of self-realization.

23. Rovelli, Carlo. The Order of Time. Riverhead Books, 2018.

- Rovelli's insights into the non-linear nature of time and its deep connection with consciousness resonate with the book's approach to time as a fluid and malleable dimension.

24. Ibn Arabi. The Meccan Revelations. Translated by William Chittick and James Morris, Pir Press, 2005.

- Ibn Arabi's mystical views on the unity of existence and the interplay between the divine and the material world subtly influence the book's metaphysical musings.

25. Goswami, Amit. The Self-Aware Universe: How Consciousness Creates the Material World. TarcherPerigee, 1995.

- Goswami's quantum consciousness theory directly informs the book's core proposition that consciousness is fundamental to the universe and shapes reality.

These references represent a constellation of ideas and philosophies that have shaped the thematic landscape of The Quantum Soul: Exploring the Metaphysics of Consciousness. While the book presents an original narrative, it is in dialogue with these broader intellectual and spiritual traditions, offering a synthesis of insights from quantum theory, mysticism, metaphysics, and consciousness studies.

GLOSSARY OF TERMS

Akashic Field: A concept borrowed from mystical and metaphysical traditions, referring to a cosmic memory or informational field that is believed to store all knowledge, experiences, and thoughts across time and space. Often described as a subtle energy that connects all things, the Akashic field is sometimes viewed as the universe's blueprint.

Anima Mundi: A Latin term meaning "the soul of the world." It refers to the concept that the universe is imbued with a living soul, which connects all matter and consciousness, suggesting that the cosmos itself is a unified, sentient being.

Atman: A Sanskrit term from Indian philosophy and spirituality, referring to the inner self or soul. In some traditions, it is considered synonymous with Brahman, the ultimate reality or universal consciousness.

Awakening: A process of heightened awareness in which an individual comes to a deeper understanding of the self, reality, and existence. Often linked with spiritual enlightenment, awakening involves the dissolution of illusions and the realization of the true nature of consciousness.

Becoming: The continuous process of change and transformation in which beings and forms evolve. In metaphysical terms, becoming refers to the dynamic unfolding of reality, where nothing remains static, and all things are in perpetual motion toward new states of existence.

Brahman: In Indian philosophy, Brahman is the ultimate, unchanging reality, beyond the reach of human perception and intellect. It is the source of all that exists, often described as formless, infinite, and eternal. Brahman is the foundation of the cosmos and is present in all things.

Collective Unconscious: A concept from Carl Jung's psychology, referring to a part of the unconscious mind shared by all human beings. It contains archetypes, symbols, and primordial images that shape human experience, representing the universal aspects of the human psyche.

Consciousness: The state of being aware of and able to think, perceive, and experience. In metaphysical terms, consciousness is considered the fundamental essence of reality, the medium through which reality is created and perceived. It is often seen as both personal and universal, transcending individual minds.

Cosmic Mind: A metaphysical concept suggesting that the universe itself possesses a mind or intelligence. This intelligence governs the laws of nature, the unfolding of events, and the interconnectedness of all beings and phenomena within the cosmos.

Cosmic Unity: The idea that all things in the universe, from the smallest particles to the largest galaxies, are interconnected and form an indivisible whole. This concept suggests that all beings and objects are bound together by a deeper, underlying unity.

Duality: The division of reality into two opposing aspects, such as light and darkness, good and evil, or body and soul. Duality is often seen as an illusion in metaphysical traditions, with ultimate reality being non-dual, transcending these apparent opposites.

Ego: The sense of self or "I" that identifies with the body, mind, and personal experiences. In many spiritual and philosophical traditions, the ego is seen as a limited and false construct, which veils the deeper, true self that is connected to universal consciousness.

Emptiness: A concept from Buddhist philosophy, particularly in the Madhyamaka tradition, referring to the inherent lack of independent existence in all things. Emptiness suggests that all phenomena are interdependent and lack a fixed, permanent essence, pointing to the illusory nature of perceived reality.

Entanglement: A phenomenon in quantum physics where particles become interconnected in such a way that the state of one particle instantaneously affects the state of another, regardless of distance. This phenomenon is often invoked in metaphysical discussions to illustrate the interconnectedness of all things.

Eternal Present: The idea that past, present, and future are illusions and that true existence occurs in a timeless state of "now." This concept is central to many spiritual teachings, which emphasise that ultimate reality and consciousness exist beyond the flow of linear time.

Field of Potentiality: A term used in quantum mechanics and metaphysics to describe the unmanifest state of reality, where all possibilities exist as potential until they are observed or actualised. It refers to the realm of infinite possibilities from which reality emerges.

Formlessness: The state of being beyond physical shape or structure. In metaphysics, formlessness refers to the ultimate nature of reality, which transcends all forms and appearances. Often associated with the spiritual or divine, it is the unmanifest source from which all forms arise.

Implicate Order: A term introduced by physicist David Bohm to describe the underlying order of the universe, where everything is interconnected in a deep, holistic way. This contrasts with the explicate order, which is the manifest, material world we experience through our senses.

Infinite: That which has no limits, boundaries, or end. In metaphysical terms, the infinite is often associated with the divine, the eternal, or the absolute nature of the cosmos, which transcends the finite constraints of time, space, and form.

Metaphysics: A branch of philosophy that explores the fundamental nature of reality, existence, and consciousness. It seeks to understand what lies beyond the physical world and the laws of nature, addressing questions of being, the nature of the universe, and the relationship between mind and matter.

Morphogenetic Fields: A concept developed by biologist Rupert Sheldrake, proposing that biological forms and behaviours are guided by fields of information or memory. These fields influence the development and evolution of organisms, shaping their structure and patterns.

Non-duality: A philosophical and spiritual concept that transcends the perception of opposites or duality. It suggests that all distinctions, such as self and other or matter and spirit, are illusions and that reality is an indivisible whole, a singular essence of being.

Omega Point: A concept proposed by Pierre Teilhard de Chardin, referring to the ultimate point of convergence for all of consciousness and evolution. It represents the final state of spiritual unity and the culmination of cosmic evolution, where the universe achieves its highest potential.

Oneness: The state of being one, unified, or whole. In metaphysical thought, oneness refers to the inherent unity of all existence, suggesting that all beings, objects, and phenomena are interconnected aspects of a singular reality or consciousness.

Quantum Field: In quantum physics, the underlying field from which all particles and forces arise. It represents the fundamental nature of reality, where particles are seen as excitations or disturbances in the field. This concept is often used in metaphysical discussions to describe the source of all matter and energy.

Quantum Soul: A concept that proposes consciousness

operates at a quantum level, suggesting that the soul or consciousness is not bound by material reality or the brain but exists as a field or state that interacts with the quantum world.

Singularity: A point in space-time where the known laws of physics break down, often associated with black holes or the Big Bang. In metaphysical discussions, singularity can also refer to the point of unity or oneness, where distinctions between self, time, and space dissolve.

Soul: The immaterial, eternal essence of a being. In metaphysical terms, the soul is often seen as the true self, beyond the mind, body, and ego, and is considered to be an expression of the universal consciousness or divine essence.

Source: The origin or fundamental essence from which all things arise. In metaphysical and spiritual teachings, the source is often equated with the divine, the absolute, or the infinite reality that lies beyond the physical world and gives rise to all existence.

Spiritual Awakening: The process of realising one's true nature beyond the ego, often accompanied by a deep sense of connection to the universe, heightened awareness, and a transformation in consciousness. It is seen as a return to the original, unconditioned state of being.

The Void: In metaphysical and spiritual traditions, the void refers to the space of pure potentiality, a formless state of being from which all things emerge. It is often associated with emptiness or nothingness, yet it is full of creative potential and the source of all creation.

Unmanifest: Referring to that which exists in a potential, unseen state before being brought into form. In metaphysical discussions, the unmanifest is the realm of pure consciousness, energy, or spirit from which the material world emerges.

Unity Consciousness: A state of awareness in which an individual experiences the interconnectedness of all things and realises the underlying oneness of existence. This level of consciousness transcends the dualistic perception of self and others, merging personal identity with the universal.

Wholeness: The state of being complete, unified, and undivided. In metaphysical thought, wholeness refers to the integration of all aspects of being, where the individual recognises their connection to the universal and the self is understood as an expression of the infinite.

ACKNOWLEDGEMENTS

The creation of 'The Quantum Soul' has been a journey of profound introspection, curiosity, and collaboration. It is with immense gratitude that I extend my heartfelt thanks to all those who have supported, inspired, and challenged me along the way.

First and foremost, I am deeply indebted to the many philosophers, scientists, spiritual teachers, and thinkers—both ancient and modern—whose works have inspired much of the reflection and inquiry in this book. Though their names are too numerous to mention individually, their contributions to human understanding are the bedrock on which this exploration stands.

To my family and friends, your unwavering support has been the grounding force throughout this endeavour. Your patience, encouragement, and belief in the vision behind this book allowed it to come to life.

Special thanks to my publisher, Irene Minds, for bringing this project to fruition and for providing invaluable guidance and editorial insight throughout the process. I am deeply appreciative of your dedication to this work and the trust you've placed in its message.

To my readers, whose curiosity and open minds bring this book into dialogue with the larger world—thank you. I hope that this exploration sparks something within you, inviting you to look deeper into the mysteries of existence, consciousness, and the boundless potential that lies within us all.

Finally, I wish to acknowledge the unseen forces—the ineffable mysteries of the universe itself—that have shaped both this book and the path that led to its creation. This work is a humble reflection of the endless journey of discovery and transformation, a journey that continues for us all.

With deep gratitude,

Dr Bhaskar Bora

COPYRIGHT INFORMATION

The Quantum Soul
Copyright © 2024 by Dr Bhaskar Bora
All rights reserved.
First published in 2024 by Irene Minds.

No part of this publication may be reproduced, distributed, or transmitted in any form or by any means, including photocopying, recording, or other electronic or mechanical methods, without the prior written permission of the author and publisher, except in the case of brief quotations embodied in critical reviews and certain other non-commercial uses permitted by copyright law. For permission requests, please contact the publisher at the address below:

Irene Minds
Contact: bora@gmail.com

DISCLAIMER

This book is a work of exploration into metaphysical and philosophical ideas about consciousness, reality, and the nature of existence. The views, interpretations, and opinions expressed herein are those of the author and are meant to inspire reflection and dialogue. They do not constitute scientific, psychological, or medical advice and should not be used as a substitute for professional consultation in these fields.

While the author has made every effort to ensure the accuracy and integrity of the information contained in this book, neither the author nor the publisher assumes any responsibility for errors, omissions, or any consequences arising from the application of the information provided. Readers are encouraged to engage with the material critically and to form their perspectives based on further study and personal experience.

The references to philosophical, spiritual, and scientific works within this book are intended to honour and draw from the wisdom of various traditions. However, this book is an original work of thought, and any similarities to other works are either coincidental or derived from sources explicitly cited.

www.ingramcontent.com/pod-product-compliance
Lightning Source LLC
Chambersburg PA
CBHW052145220526
45471CB00004B/1526